"The search for understanding the secrets of our existence and the secrets of the universe is as old as the age of mankind. However, with the emergence of the scientific methods we advance enormously from believing in arbitrary assumptions and superstitions to having a reliable way to explore how nature works. This also means that anyone has the ability to look at major puzzles and to even show that extremely prominent scientist views are wrong. Fortunately, now you can do so without the fear of being prosecuted.

Solving problems in science is like talking with angels, but in this case knowing that you are not dreaming. One may think that Science has very little to discover in the present time, but this is clearly not the case. Scientists have discovered some amazing facts about nature though the greatest puzzles remain to be solved. These include the molecular origin of our emotions, the origin of the universe and less challenging puzzles like how to extend our life and how to generate endless energy. The satisfaction of even trying to solve any great scientific problem is one of the most fulfilling tasks."

Arieh Warshel
Nobel laureate
Member, National Academy of Sciences, HonFRSC
Dana and David Dornsife Chair in Chemistry
Distinguished Professor of Chemistry and Biochemistry
Member, USC Norris Comprehensive Cancer Center
USC Department of Chemistry

"One way of appreciating science is by imagining a situation where there was no science. Without science, how would we have looked at the world around us. Because of the discoveries, inventions and innovations, science has contributed to better the quality of life for human beings. While it has also made contributions some of which may not be good for human beings when wrongly applied, the most important feature of science is that it gives us a way of understanding properties and phenomena and a way of questioning. It has great cultural value as well, and has emerged to become part of our culture. Those who feel that science is highly technical with no relationship to other activities of humans, are missing the point. The beauty of science is that it not only looks at the future, but also at the present. Scientific attitude or scientific temper is an essential quality that distinguishes a concerned citizen. For those who work in science, science can give a way of life. It may not be an exaggeration if I were to state that science in the broadest sense can ensure peace and harmony in the world at large, and help eliminate narrow dogma.

Lest one forgets, science gives pure, distilled knowledge. It is for us humans to use it judiciously and with purpose.

Maybe it is because of science, that I am one of the happiest persons that I know of, doing what I enjoy most, and being positive about life and humanity as a whole."

C.N.R. Rao, F.R.S.
Honorary President & Linus Pauling Research Professor
Jawaharlal Nehru Centre for Advanced Scientific Research

"People love adventure stories — stories of mystery, where detectives solve an age-old secret, or explorers discover a long-lost treasure. I think of science as an adventure story. Nature has hidden secrets all throughout the universe, and it's our job to crack the code and solve the mystery. What could be more exciting? Except that these mysteries are about existence itself: How did the universe begin? What is life? How does the brain compute our thoughts and feelings? What could be better than an adventure story where your goal is to crack an age old mystery about the universe itself! And solving these mysteries has an interesting byproduct effect. If you solve a mystery, then you sometimes can have great impact on everyday life and everyday happiness. Think of how drugs that treat infections, airplanes that soar through the air, computers that can predict the weather — these and countless other products of successful scientific adventures have changed our ability to thrive, connect, and flourish. What could be better than that — a mystery and adventure story that, when solved, can make the world a better place?"

Ed Boyden, Ph.D.
Y. Eva Tan Professor in Neurotechnology at MIT
Leader, Synthetic Neurobiology Group
Associate Professor, MIT Media Lab and McGovern Institute,
Departments of Biological Engineering and Brain
and Cognitive Sciences
Co-Director, MIT Center for Neurobiological Engineering
Member, MIT Center for Environmental Health Sciences,
Computational and
Systems Biology Initiative, and Koch Institute
Massachusetts Institute of Technology

"Pure scientific inquiry and its currency, discovery, are the greatest gifts of humanity. A scientific discovery has the potential of transforming us as human beings by transforming our understanding of each other and the world around us. Consider how the language of science allows friends and foes to speak and exchange ideas without regard to who they are and where they come from. It is the greatest equalizing force on earth for its freedom in inquiry and expression. How does this actually happen? It starts by going into the unknown guided by established facts, with some passion, and the unshakable belief that Nature will reveal itself through a scientific experiment. An experiment done for the sake of advancing human knowledge, with no influence from a boss, no influence from a societal need, no influence from a funding agency, no influence from the experimenter's own bias, gives the best chance for the most beautiful and impactful discovery to be made. Such discovery having been made away from the limelight and the hurly-burly of daily life has the potential of leaping to the forefront of solving society's vexing problems. After all, as scientists, we represent a tiny fraction of society and it is us who are asked to solve society's problems of disease, drought, pollution and the like, don't we deserve the solace of our experiment? I am fortunate in my life to have the freedom to do the unbiased experiment, and to work with students who in the trenches of the experiment 'fail, fail and fail' so that they may succeed, discover, and help build a better world."

Omar M. Yaghi
James and Neeltje Tretter Chair Professor of Chemistry
University of California, Berkeley
Faculty Scientist
Lawrence Berkeley National Laboratory
Co-Director
Kavli Energy NanoSciences Institute at Berkeley
California Research Alliance by BASF

The Promise of Science

of Science

**Essays and Lectures
from Modern Scientific
Pioneers**

Other Related Titles from World Scientific

The Promise of Science

of Science

Essays and Lectures from Modern Scientific Pioneers

Editor

Lorie Karnath

Molecular Frontiers Journal, Molecular Frontiers Foundation

World Scientific

NEW JERSEY · LONDON · SINGAPORE · BEIJING · SHANGHAI · HONG KONG · TAIPEI · CHENNAI · TOKYO

Published by

World Scientific Publishing Co. Pte. Ltd.

5 Toh Tuck Link, Singapore 596224

USA office: 27 Warren Street, Suite 401-402, Hackensack, NJ 07601

UK office: 57 Shelton Street, Covent Garden, London WC2H 9HE

Library of Congress Cataloging-in-Publication Data

Names: Karnath, Lorie, editor.

Title: The promise of science : essays and lectures from modern scientific pioneers /
    edited by Lorie Karnath (Jason Learning, Germany).

Description: New Jersey : World Scientific, 2018. | Includes bibliographical references.

Identifiers: LCCN 2018045374 | ISBN 9789813273283 (hardcover : alk. paper)

Subjects: LCSH: Science--Miscellanea. | Science--Popular works.

Classification: LCC Q173 .P96275 | DDC 500--dc23

LC record available at https://lccn.loc.gov/2018045374

British Library Cataloguing-in-Publication Data

A catalogue record for this book is available from the British Library.

For any available supplementary material, please visit
https://www.worldscientific.com/worldscibooks/10.1142/11071#t=suppl

Typeset by Stallion Press

Email: enquiries@stallionpress.com

Introduction

This book offers a broad selection of thought-provoking discoveries and astonishing cutting edge technologies, authored by some of today's most renowned scientists. Many of these are addressing, and even solving, challenges oftentimes before they have been even recognized as issues. Herein lies the promise of science, to unlock groundbreaking new potentials that can serve to expand and improve the world and all that surrounds it. Through imagination, insight, research and discovery, science continues to provide pathways that offer means to move forward in ways most could never have dreamed of.

Without doubt, today science and its technologies are transforming civilization as never before. While the desire to unravel the mysteries of the Universe and life itself have been inherent since the beginning of mankind, relatively recent and ongoing developments which have produced increasingly sophisticated and reliable tools and methodologies have allowed scientists to delve deeper and more effectively than ever before into their desired areas of research, unravelling puzzles previously considered insurmountable. As science continues to offer an increasingly dazzling array of opportunities, it is crucial to ensure that those willing are provided the potential means to be engaged in the process. Continuing in this manner, ensuring that new developments with universal validity and utilization are embraced, rather than rejected or ignored due to unnecessary lack of understanding or fear. Imagine, Explore, Discover, Share — these are the words that science must live by. It is through exploration and its resultant knowledge, together with the communication of that which has been newly revealed, that leads to discovery and acceptance.

Today science and technological developments are the drivers behind transformations both big and small, integral to our lives in a myriad of fundamental ways that enhance our daily existence and future potential. The callings of science these days in our global world are often far reaching, rarely within the realm of the individual but rather representing a tremendous pool of talent, disciplines and resources. A breakthrough in one arena may lead to far reaching consequences and applications elsewhere, meaning that science does not fail or flourish in isolation. It represents an ongoing interaction and exchange of ideas fuelled by imagination and accumulated experience. It is this amalgamation of knowledge, efforts and resources that brings forth discovery, gradually unlocking the many secrets that life, the planet and the space that surrounds us present.

To truly fulfill its promise, science requires the willingness to venture not only beyond recognized boundaries of unknown territory, it must exceed the limits of established convention, thought and knowledge, requiring an original perspective on what was once considered familiar terrain. In the process the scientist oftentimes will open the door to pioneering thoughts providing new input as to our perception of our world, in the process redefining our place within it.

Lorie Karnath
Bengt Nordén

About Lorie Karnath

Lorie Karnath is the managing editor of the *Molecular Frontiers Journal*, an open access publication dedicated to the sciences. She is also a founding member and director of the Molecular Frontiers Foundation — a not-for-profit organization that seeks to inspire interest in the sciences.

Lorie has authored numerous books on the sciences, discovery and exploration, including a children's book co-authored with Gustav Born that answers the questions that young people ask in simple scientific terms. She served as the president of the renowned Explorers Club and is an active board member of a number of scientific and educational organizations. She has an MBA from INSEAD and an honorary Ph.D. from Shenandoah University.

About Bengt Nordén

Professor Bengt Nordén is a physical chemist with interest in structure and interactions in biomolecular solution systems. His pioneering contributions cover a range of physics and chemistry: from fundamental optical properties of bio-chromophores to topics related to molecular recognition.

Prof Nordén is a former Chair of the Nobel Prize Committee for Chemistry. Based on his international networks he founded the Molecular Frontiers Foundation (MFF, www.molecularfrontiers.com) hosted by the Royal Swedish Academy of Sciences. MFF has as purposes to identify early on scientific breakthroughs and to empower youth in demonstrating the usefulness of science. He also serves as the Editor-in-Chief of the *Molecular Frontiers Journal*.

Thoughts About Science

"Science, like nothing else among the institutions of mankind, grows like a weed...

Science consistently produces a new crop of miraculous truths and dazzling devices every year,

truths and devices that enrich our lives and grow up out of the graciously willing puzzles of the unknown in an orderly but unpredictable way,

out of a process of observation, hypothesis, experiment, conclusion; a process that as far as we know,

was first proposed and adopted, only a few hundred years ago by a number of Europeans faced with a new world to explore

and some worn out scholastic tools passed down from the ancient Greeks to explore it with.

"The Galileos and the Newtons and the Hookes, and the Boyles invented new sharper tools of inquiry and the age of science was born.

Now we each of us have things and thoughts and descriptions of an amazing universe in our possession

that kings in the Seventeenth Century would have gone to war to possess.

We are the recipients of scientific method. We not only can luxuriate in its weed-like growth,

but we can each of us be a creative and active part of it if we so desire. And we will.

There is no stopping it, nor can there be any end to it."

Kary Mullis
1993 *Nobel Laureate in Chemistry*
Inventor of the Polymerase Chain Technique

Contents

Editorial — The Knowledge Gap

Gustav V.R. Born and Lorie Karnath

Science and its technologies, with their universal validity and utilization, should bring people together more effectively than any other human activity. Indeed, they do this brilliantly for those devoted to common research pursuits, such as the worldwide collaborations in genomics and proteomics, the much larger task of finding out how proteins do their work in cells; and for numerous commercial technologies, such as satellite communications.

But the ever-accelerating acquisition of knowledge has also had the opposite effect, through increasing the separation of those who are part of this process from those who are not. Such a gap has of course existed since modern science began in the 17th century; but by now it has become a schism between different mental worlds. This causes misunderstandings, antagonisms and confrontations.

---

[1] FRCP, FRS, Research Director, Emeritus Professor of Pharmacology at King's College London and Research Professor at the William Harvey Research Institute, Barts and The London School of Medicine and Dentistry.

[2] (hon) FRGS, Molecular Frontiers Foundation, chair SAB, Managing Editor *Molecular Frontiers Journal*.

About Gustav V.R. Born

Gustav Victor Rudolf Born FRCP, HonFRCS, FRS (29 July 1921–16 April 2018) born in Germany, the son of Hedwig Ehrenberg and the scientist Max Born. His family left Germany, as his father and maternal grandfather were Jewish. He was educated at the Oberrealschule, Göttingen. After moving to the UK with his father, he attended The Perse School, Cambridge, Edinburgh Academy and the University of Edinburgh.

From 1973 to 1978 he was Sheild Professor of Pharmacology at Cambridge. He was elected a Fellow of the Royal Society of London in 1972, and of the Royal College of Physicians in 1976. He was Professor of Pharmacology at King's College London, 1978–86, and became Research Professor at the William Harvey Institute in 1989.

The knowledge gap does not preclude the ability of everyone on both sides of the divide to make use of the most sophisticated scientific technologies. It is amazing how the millions who talk to each other around the world on mobile phones rarely if ever look at the little gadget with the awe it deserves. Numberless essential activities are based on scientific knowledge without needing explicit understanding. Technicians in medical laboratories know how to determine the presence or absence of each of the many proteins essential for blood clotting, without understanding how they bring this about. To do such work effectively requires, as in innumerable other activities, knowledge up to a certain level, so that imparting the knowledge-base of all imaginable skills is an important element in calls for "education, education, education".

But looked at right across the world this level of education remains way behind what the technological environment requires. Poor countries struggle to maintain the even more basic learning needed for survival. In many countries education is impeded by bigotry, as when a total ban on women's education was imposed in Afghanistan, and as in some parts of the United States where teaching of the theory of evolution is forbidden — This at a time when the President of the Royal Society can speak of Darwin's theory as having much the same standing as Newton's law of gravitation. Even those going up to University from the best and most expensive private schools may be thoroughly conversant within the humanities while lacking all knowledge of basic biology or even of their own body functions, thus reinforcing C.P. Snow's "two cultures".

Thus, whilst just about everybody uses scientific technologies, only a small proportion understand or want to understand what they are using. Many people lack the time, the energy or the willingness to take in new kinds of information and to think in unfamiliar ways. To the extent that remains true, this knowledge gap persists.

But knowledge is no more than a tool in the quest for understanding. Understanding the mechanisms of blood clotting may be of no direct concern to laboratory technicians but is crucial for devising antithrombotic drugs. This understanding came about when Gwyn Macfarlane in Oxford and Oscar Ratnoff in Cleveland independently proposed that each of the many essential proteins acts on the next in a sequence or cascade, ending with the formation of the insoluble protein fibrin that makes the clot. The general point is that the juggling of quite a limited number of facts in receptive scientific minds can bring forth far-reaching generalizations. That has been the essence of scientific discovery, and shows up the gap between knowledge and understanding.

In many fields of science this *modus operandi* is in danger of being buried under avalanches of new knowledge. From astrophysics to genetics and proteomics, technological developments in combination with computerization are generating factual information at much greater rates than individuals or even super-computers can cope with. This has created a new kind of gap within the scientific community itself, by producing opposing effects on the discovery scene.

On the one hand, it is undermining the classical road to discovery. This was put starkly by the Research Director of a renowned pharmaceutical company: "We are no longer interested in new drug ideas. Our entire effort is concentrated on screening hundreds of thousands of compounds for potential therapeutic activities." In *The Future of Life* Edward Wilson makes the same point, namely that serendipity is the hallmark of pharmacological research; he should have inserted the word "now". It seems that approaches based on knowledge and understanding of disease are going out of favor. When blind technology ousts understanding to that extent, something of proven worth must be getting lost. Presumably there will be no follow-up flashes of insight such as the moment when Hans Kosterlitz, after years of hard-won knowledge about morphine, asked himself why the poppy should produce such a potent pain killer: might a substance with similar actions be present in the brain itself? This simple question opened up a new and fruitful field of knowledge about pain relief, something of great importance to every human being.

On the other hand, none of this is to deny the potential of serendipity. Edward Wilson gives examples in support of his assertion that chance discovery can lead not only to a successful drug but also to advances in fundamental science, which in time yield other successful drugs. One of Wilson's examples is a synthetic drug more effective than opium as a pain killer but non-addictive, a discovery based on screening of the toxins produced by a Columbian poison dart frog. As this work is in the same field as that of Kosterlitz, mentioned earlier, it brings out the contrast very clearly. Another example concerns the discovery of cyclosporin, a powerful suppressor of the human immune system, used routinely to prevent rejection of transplanted organs, which was discovered during routine screening from an obscure Norwegian fungus.

What all this tells us is that much of science is becoming ever more a process in which a discovery is expected to emerge from a multitude of interactions rather than being the clear-cut result of one person's ideas and experiments. This development is making itself felt even in the Nobel prize awards. According to Alfred Nobel's will, the prizes for physics, chemistry and physiology or medicine cannot be shared by more than three persons. Nowadays the prize is usually so divided, presumably due to the oftentimes collaborative nature of discovery.

In principle it should remain possible to fish new ideas out of the oceans of information, and no doubt this is happening. But many scientists feel obliged to keep up with new technologies and new results in their fields; and they may thereby be practically, "no time", and psychologically, "all too much", hindered from what should always be their core activities — thinking and imagining.

Thus there is a gap within the scientific community between what might be called the "automaters" and the "cerebraters".

Presumably both are needed for progress from now on. Actually it is conceivable that even their joint endeavors may hit an ultimate barrier to understanding nature's greatest complexities, such as the cosmology conundrums, the interdependence of species, and the human brain; as J.B.S. Haldane said: "The world may not only be queerer than we think but queerer than we *can* think."

For their part the general public, by no means well informed about science in the first place, is faced with ever more limited access to new knowledge by barriers of difficult technologies and terminologies. At the same time people are bombarded by journalistic announcements of amazing discoveries and new therapies well before they are confirmed. The capabilities of new technology and the significance of new knowledge are difficult to make intelligible. It is not surprising therefore that the public becomes indifferent to all this newness and discouraged about science and what it stands for. *Pari passu* people become more susceptible to the blandishments of all kinds of unscientific nonsense, most dangerously to realms of pseudo-medical nonsense, which may preempt proper diagnosis and treatment of serious disease.

Well-devised efforts to combat misinformation and to get the scientific basis of modern life across to the general public are made by a number of bodies such as the Committee for the Public Understanding of Science (COPUS) of the British Royal Society: but the scale of the task limits success. The gap between the scientific know-hows and the know-nots is actually being widened by the progress of science, presenting an increasing obstacle to global cooperation and so becomes a cause for anxiety. Importantly as science continues to grow and flourish we need to work harder to ensure its more widespread and inclusive dissemination. Therein lies the potential and promise of science.

This issue of *Molecular Frontiers Journal* highlights a number of the wide-ranging means and methods in which technologies are impacting scientific discovery. While not a new phenomenon as science and technology have maintained a symbiotic relationship since early times, the breadth and scope of which advances are taking place due to the myriad of ways that these are connected, seem to be ever growing. In some instances, it is a new scientific discovery that fosters a new technological direction, in others, it is the technological innovation which may facilitate fresh scientific inspiration. Technology has also served to amplify the role of science well beyond the laboratory, increasingly scientific endeavors are focused with consideration towards projecting what might and could happen and how to prepare for this. Even the scientific protocols and practices themselves applicable today can no longer be presumed as relevant going forward.

Undoubtedly, emerging technologies and methods of utilizing these are providing for novel directions, procedures and applications in the sciences. The speed at which new platforms are evolving assure that there will be profound changes ahead. In tangent with technological advances and uses, innovation is taking on an expanded and forward looking context in trying to predict how such technologies might be used and how science prepares for these applications in the future. Technological breakthroughs are providing platforms for science to think in entirely different contexts. At present scientists are only beginning to imagine how research, and the dissemination, sharing and application of results will evolve based on the ability to interact in a three-dimensional manner, to be able walk around or through something in an interactive manner. In addition to how increasingly sophisticated capabilities such as those that allow for the acceleration or for complete or partial virtual reactions at varying stages and distances while providing the capability to document irrevocably the source of contributions at each interval. What role artificial intelligence may play in the sciences and research arenas are also still to be defined.

Keeping up with the ever-evolving world of scientific pursuit and the discoveries that this brings has always been difficult, however now due to the flurry of revolutionary technologies this has become perhaps even more-so. Certainly, the inequalities between those with technological access and those who do not are further exacerbated. As scientists grapple with the plethora of new technological prospects ahead, we must work ever harder to ensure that the messages of science and the technological and other tools that disseminate these are available to all.

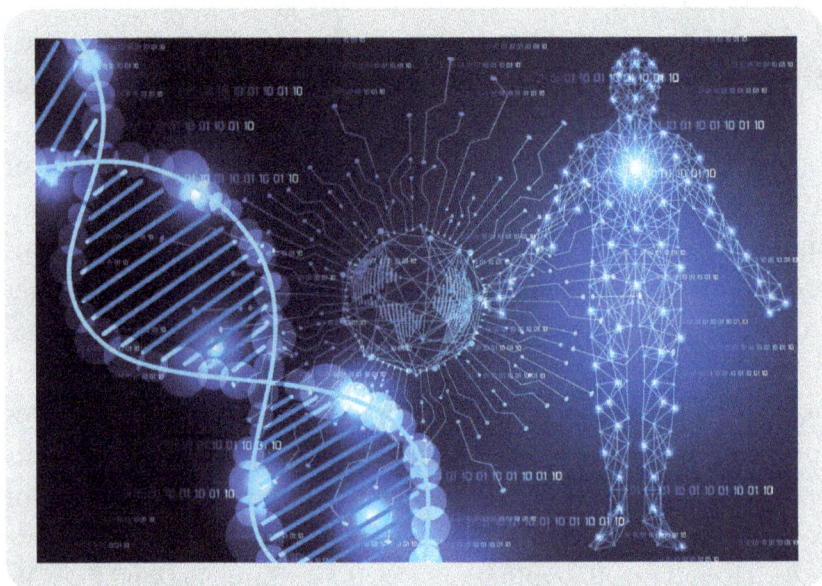

Nanoscience — Potential and Threats*

Paul Alivisatos[1]

As the world's population continues to expand, scientists are working to address the energy needs and challenges that accompany growth with environmentally responsible approaches. Nanoscience is helping to provide solutions to energy and environmental concerns in a number of ways.

Keywords: Nanoscience; Energy; Environment; Carbon Cycle 2.0 Initiative.

---

* This chapter was based on a lecture delivered by Professor Alivisatos at the Molecular Frontiers Symposium in the Royal Swedish Academy of Sciences in May 2012.
[1] Department of Chemistry, UC Berkeley, Berkeley, CA 94720-1460, USA.

About Paul Alivisatos

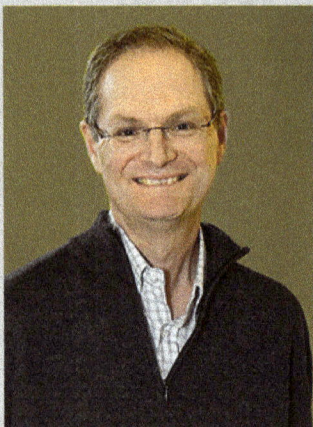

Armand Paul Alivisatos is the University of California (UC) Berkeley's Executive Vice Chancellor and Provost and Samsung Distinguished Professor of Nanoscience and Nanotechnology. He is the Director Emeritus of Lawrence Berkeley National Laboratory, which he was the Director of at the time this lecture was delivered. He is a founder of two prominent nanotechnology companies, Nanosys and Quantum Dot Corp, now a part of Life Tech.

Groundbreaking contributions to the fundamental physical chemistry of nanocrystals are the hallmarks of Alivisatos' distinguished career. He has demonstrated key applications of nanocrystals in biological imaging, electronic display technologies, and renewable energy.

Alivisatos has been recognized for his accomplishments, with awards such as the Dan David Prize, the US National Medal of Science, and the Wolf Prize in Chemistry.

Alivisatos received a Bachelor's degree in Chemistry in 1981 from the University of Chicago and Ph.D. in Chemistry from UC Berkeley in 1986. He began his career with UC Berkeley in 1988 and with Berkeley Lab in 1991.

It's really fun to be part of a Molecular Frontiers Symposium. This is where you can think about deeper questions that maybe come up in your own discipline each day by joining with others and of course to interact with people who are early in their careers and try to help them find where they may be able to contribute, and to learn from them as well. The title of my talk which was given to me, seeks an explanation of nanoscience, both its advantages and its potential problems, and to do this within the context of the energy and environment problem. So this is not an easy task and, and we will have to move kind of quickly here to go through some aspects of what is nanoscience in a short period of time.

In fact it puts me in mind of a famous TV character that used to be on the television in the United States called Father Guido Sarducci. He had a funny comedy routine once which was called "The Five-Minute University." In the Five-Minute University you learned in five minutes everything that you would still remember five years after you attended a full course at a university. This is going to be our Five-Minute University version of what is nanoscience but hopefully in the context of energy and environment.

And I think that you can appreciate that in coming decades, problems of energy and environment will really intensify, as the 10 billion people we've been talking about will represent a significant increase in the population. But, as well they will be using energy with greater energy intensity because all of the people who are today in the low economic stratum will be pulling themselves up every way they can in order for their families to be better off. And so we will have greater and greater energy use by a larger number of people and therefore we know that there's going to be a big crunch coming. And the question is how can we also organize ourselves, as a science community, to meet some of those demands. And I wear another hat as a laboratory director with a laboratory that has a few thousand people working at it, so many of them on energy and environment problems, and there too we have to think how we can organize ourselves. We also know that any solution that we come up with in order to be important actually has to operate on a massive scale and therefore it could have unintended consequences, and we need to have a system for learning how to think about the consequences of our technologies and the unintended consequences as early as we can in order to be able to be proactive.

So to give it some context, today we operate in our energy world with an open-ended carbon cycle. That means that we burn fossil fuel and put CO_2 into the atmosphere. About half of the molecules that we put into the atmosphere, 30 gigatons of CO_2 per year, about half of that stays in the atmosphere changing the radiative balance and the heating of the planet, and about half of it enters into the biosphere into the oceans and into the land masses causing various changes there. It's important to know that, that this number of sort

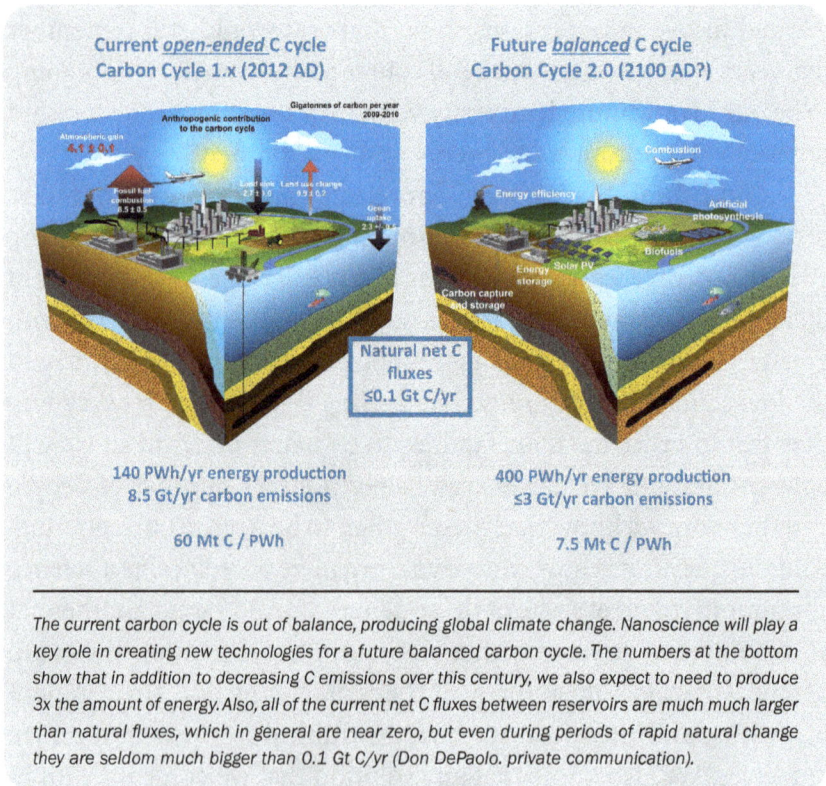

The current carbon cycle is out of balance, producing global climate change. Nanoscience will play a key role in creating new technologies for a future balanced carbon cycle. The numbers at the bottom show that in addition to decreasing C emissions over this century, we also expect to need to produce 3x the amount of energy. Also, all of the current net C fluxes between reservoirs are much much larger than natural fluxes, which in general are near zero, but even during periods of rapid natural change they are seldom much bigger than 0.1 Gt C/yr (Don DePaolo. private communication).

of eight and a half gigatons of carbon or 30 gigatons or so of CO_2 is massively larger than the net carbon fluxes that normally occur in the planet when people are not putting so much carbon into the atmosphere. The net fluxes in any given period of time in past history were much, much smaller. And so this is a large perturbation. And as we imagine moving towards year 2100 perhaps then we will have created a new, balanced carbon cycle and to get towards that we will need to develop a variety of new approaches. But especially we will have to find a way to very significantly increase our energy production and at the same time very significantly decrease the carbon emissions. So the intensity of carbon use has to go up by at least one order of magnitude. So it is very challenging.

And so at the lab, at Berkeley Lab, we've organized ourselves into what we call The Carbon Cycle 2.0 Initiative. The goal being to try to provide the science that's needed in order to have a more stable carbon cycle in the future and it's organized around each one of these circles that you see here and they are laid out in this way. Along this

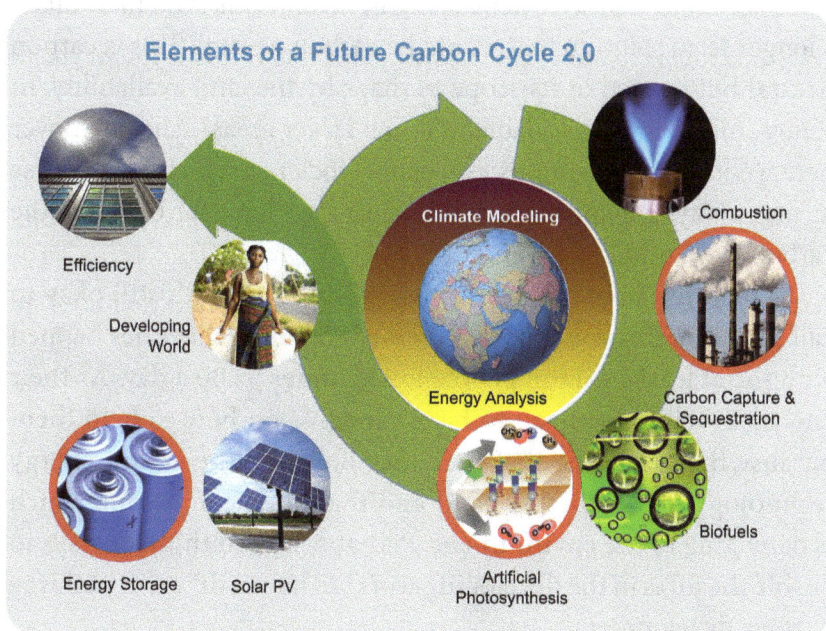

Elements of a Future Carbon Cycle 2.0

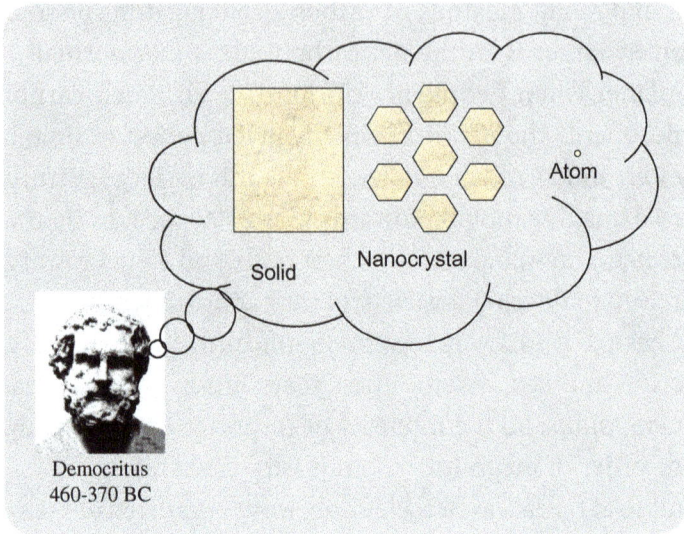

line here you can see combustion, carbon capture and sequestration, biofuels and artificial photosynthesis. You can think of this is the arc of the carbon cycle. This is how we use carbon today. And these are some solutions that we will be using in the interim. And then this is a longer term solution that would be carbon neutral. This is carbon neutral but limited in its scope perhaps by the land availability. In future, might this be scalable to much larger areas? And then over here we have the off-ramps especially efficiency which is of course in many ways the most accessible means we have for bringing the carbon cycle under control today.

I mention here the developing world, I hope it's still okay to call it that. In this context, it's intended to mean what does happen with the at least 2 billion people living under $2.00 a day. As their energy develops we need to have thoughts on how to help them because the first thing that they may do is use the easiest energy technology that comes to hand and that often can be one which is damaging to the environment. Perhaps, we can think of ways to enable the folks in the developing world to have a different pathway to their development.

And then of course there are these two off-ramps as well which take us off of carbon energy storage and solar PV. In the center sit our scientists who do climate modeling and especially those who do energy analysis and we'll see more of these towards the end. Energy analysis is what's going to help us think about all of these as well as the life cycle assessments and some of the risk assessments that we'll be talking about.

So today I'm going to talk about these three subtopics and the question that we're going to ask is, how can we have some positive outcomes here? We don't want to just be thinking about the environment as being the sort of horrible bad nightmare about what's going to be in the future. In fact actually it's a very exciting period of time because we have the opportunity to think about new ways, new and exciting problems that will enable us to have energy in a more friendly environmental way that can really advance the state of our thinking and of our science. So we're going to ask how can nanoscience specifically contribute solutions in areas of energy and environment and I'm going to give these three examples, one each in batteries, artificial photosynthesis and carbon capture and sequestration. So first of all I'm going to back up and say what are some themes of nanoscience and here are some; scaling laws, synthesis and properties of building blocks, the power of one, a deep connection to biology and the current era of building. Hopefully, by the end of the talk you'll see what each one of those means.

Now let's back up and think about the basis of nanoscience. We're going to take a slight break here, and just go back to the beginning of nanoscience which in this case we're going to attribute to Democritus in 400 or so B.C. He was a philosopher who was the first person to use the word atom. Atom means "that which cannot be divided," that is literally meaning of the word atom. And what was Democritus thinking about? Well, he was simply trying to understand the nature of matter. He was trying to think, what is stuff? And he thought okay, in order to understand it he would suggest a thought experiment. If I followed the instructions of his thought experiment, I would start with an ordinary piece of gold. I would take the gold and just chop it into two pieces and ask the question do I now still have two pieces of "gold?" Are they the same as the one big piece that I started with or not? And the fact is that for all intended purposes if you were to measure any fundamental property like say the melting temperature, the hardness, any of those kinds of properties, and you break the gold into two, the answer would be yes, they're the same as long as they're macroscopic. But somehow Democritus was able to intuit the idea that if you keep breaking it in two, break in two again and again and again, eventually you would get to a point where when you break the gold up it would now be different, matter would be grainy, and you would reach a point where you could not divide it any further. And that's the indivisible, that's the atom.

And it turns out what is nanoscience? Nanoscience is the study of matter that's in between matter where it doesn't matter how big it is anymore, whether it's twice or half the size all the fundamental properties are the same, and matter where it's truly atomic and it cannot be divided anymore. In between those two, there are a lot of sizes that are very small, you can't really see them with your eye, but every time you make the material a slightly different size, all of its properties change. And that is what nanoscience is about is to try to study matter in that very tiny regime because of course it's so interesting that you can change the properties by changing the number of atoms. So Democritus really set the stage for the study of scaling laws of nanomaterials. Scaling laws will describe how you smoothly go from here to there in properties as you change the size, in the event that it is a smooth variation.

So we're going to see some scaling laws in action and then from seeing those scaling laws, we're going to be able to make conclusions about energy technologies and that's going to be the goal of this lecture. We're going to look at four scaling laws; melting temperature, structure, energy level spacing and the concept of "artificial atoms," and charging energy. And these are four very powerful scaling laws and they relate of course going from structure to function. What we're going to find is all these properties depend critically upon the size and therefore control of size, shape, topology and connectivity on the nanoscale can enable new functional materials for energy and environment because we can control the properties by changing the sizes. We can make new materials that could have many of the properties that we want them to have.

So let's go through and see one example of a scaling law and let's start with melting temperature, I mentioned that earlier. Imagine a solid and if you think about the solid, all the atoms are holding each other together. They're all bonded with each other. So if you have a big solid, it has a "cohesive energy." How much energy it takes per atom in order to start to really get it to melt is going to depend on the size when the crystal is very small because all of the atoms on the surface are not bonded to any on the other side. So the total bonding of that

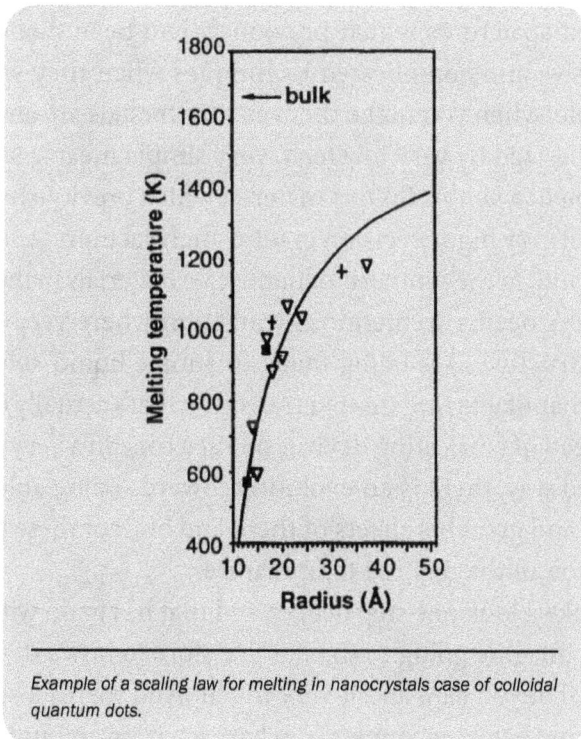

Example of a scaling law for melting in nanocrystals case of colloidal quantum dots.

crystal will become weaker when the crystal becomes very, very tiny because of all the surface atoms. And we'll see that here in action.

This is an example of the melting temperature of a solid. In this case it's a semiconductor as a function of its size. And what you can see is that as the crystal becomes smaller and smaller the melting temperature is dramatically reduced. In this case the nanometers scale bar represents 40 angstroms, or 4 nm. As we go to smaller and smaller sizes you can see that the melting temperature drops off. And there's a scaling law for this which goes roughly like one over the radius which is the ratio of the surface to the volume. In other words if there's more surface, then the melting temperature is reduced, the crystal is held together not as tightly when it becomes very, very small. Hopefully, that's reasonably intuitive.

Now the consequence of that for materials is that we can make high-quality crystals under very simple cheap conditions and that's going to be quite important. So one thing that we've seen in recent years is that structures which previously had been made by very, very intensive and complicated techniques when they were made in large scale, when you make them as tiny crystals are amenable to being synthesized by very low tech, very simple means, in this case just growing in a simple fluid, a material which previously had to be controlled at very high precision in ultra-high vacuum. And so in the research world, in the synthesis of nanoscale materials in the research world, there's been a tremendous transition where very advanced quantum structures are being made in simple liquid solutions. In industrial manufacturing this is a transition that's actually occurring where instead of fabricating devices one at a time in a very serial and complicated way, there is an evolution towards being able to make large areas and great big sheets of them and both of these flow from the reduction in the melting temperature.

Let's take a look one step deeper and that has to do with defects. If you can, imagine going to the jewelry store to buy a diamond for somebody that you care about very much. And there's a scaling law that you immediately encounter. You have a certain amount of money.

And it turns out you cannot double or quadruple the amount of money very easily and therefore, the price of the diamond, how it depends on the size really matters to you. And if you go and look at the De Beers website where the diamonds are sold, they'll tell you right away big diamonds are much rarer. So a diamond of double the weight costs four times more. That's an economic scaling law of price with volume. But why is the bigger diamond that's perfect so much more rare? Well, it has to do with the time scale for annealing a defect out. Imagine that there's a defect inside a crystal and you ask the question how long is it going to take to move out of the crystal if I warm the crystal up? Eventually, the defect will go out, it will leave, it will be pushed out. How long will that take? It turns out it will scale exponentially with the volume. And so this is why a big diamond will be so much rarer.

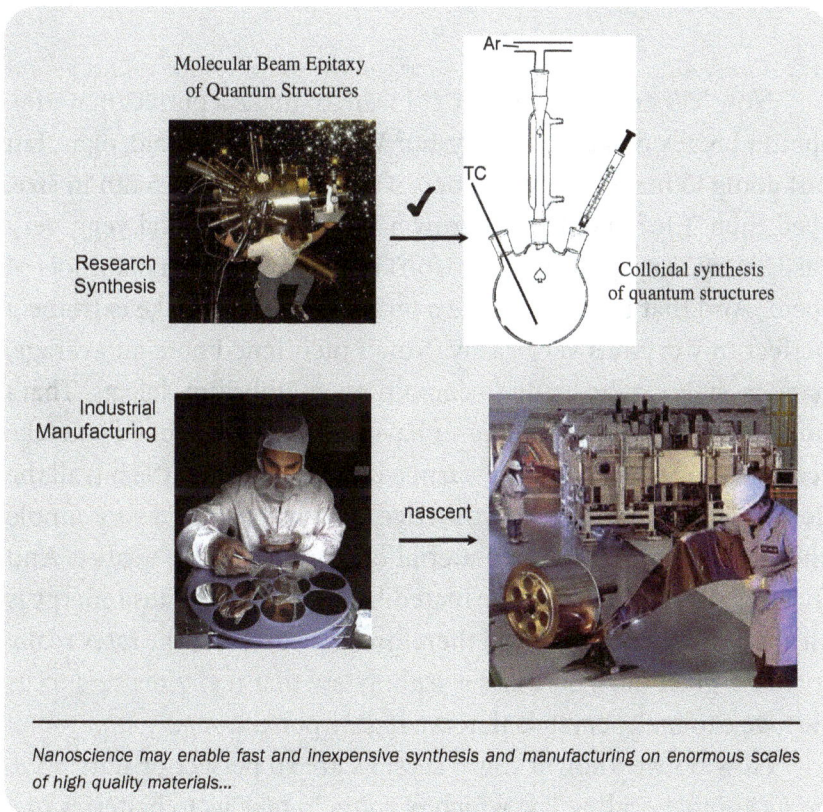

Molecular Beam Epitaxy
of Quantum Structures

Ar

TC

Research
Synthesis

Colloidal synthesis
of quantum structures

Industrial
Manufacturing

nascent

Nanoscience may enable fast and inexpensive synthesis and manufacturing on enormous scales of high quality materials...

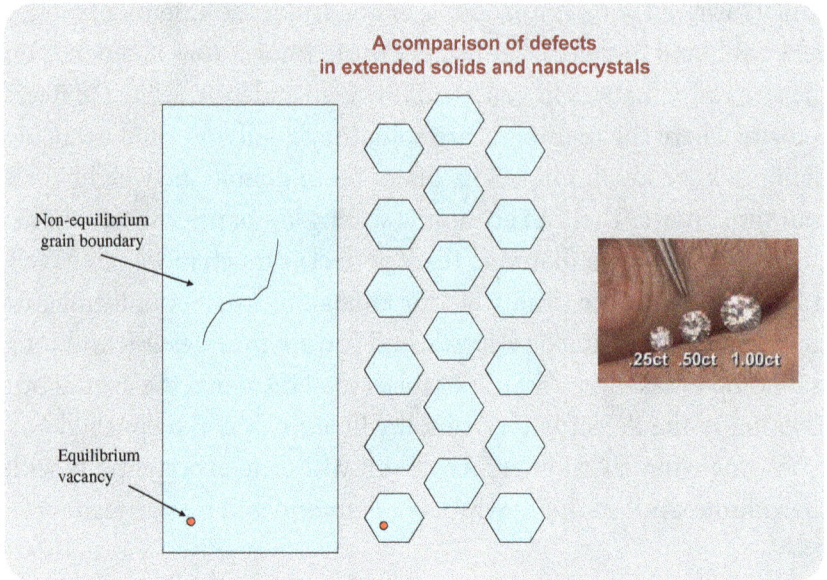

A comparison of defects
in extended solids and nanocrystals

Non-equilibrium
grain boundary

Equilibrium
vacancy

.25ct .50ct 1.00ct

Now let's turn that around. If I want to make a perfect material, should I really make a giant crystal? What if instead I said, okay, I'm just going to make a very tiny one, it's only going to be 5 nm in size. Well then it turns out I can grow a very perfect crystal very, very easily, very, very quickly and it won't be rare at all. I can make lots of them. And that turns out to be a fact that you can make extremely perfect tiny crystals very easily. Now I mentioned here on average, nanocrystals can be made to contain no equilibrium defects. That's not to say that nanocrystals never have defects in them but if you take reasonable care and you just warm a tiny crystal up a little bit, all the defects will be pushed out to the edge. That means that a very simple cheap process can make a material of extremely high quality. And that for energy turns out to be incredibly important because energy is all about huge volumes. And therefore, the costs and the rates really matter a great deal. So that's a scaling law that really matters to us that we can make crystals that are highly perfect.

How do we know if these crystals are so perfect or not? Well, here's another scaling law which is going to take us to batteries in a moment. Imagine taking a crystal, this is one shown here, where each atom is bonded to four others. And we'll simply press the heck out of it. You just push it on all sides. And the crystal now would like to become smaller because it's being pushed from every direction and so it's trying to compact itself. And what will happen is the volume will decrease as you increase the pressure. The volume will decrease and then all of a sudden at a certain pressure "boom" there'll be a big volume change and it will go from one atom bonded to four others to one atom bonded to six others. Now that can be studied in all kinds of materials but in a nanocrystal something very special happens which is that this change, which corresponds to an 18% change in the volume, it's an enormous change in the volume. This change is completely reversible in a tiny crystal. If you now release the pressure, it will pop back. You can pop between one structure and the other, despite the 18% volume change without the crystal ever having a new dislocation or a new defect generated. If you take a big crystal and you press it like this to the point where it undergoes a phase transition of that volume change, the crystal will break up into lots of little domains, but a tiny crystal can just pop between one structure and

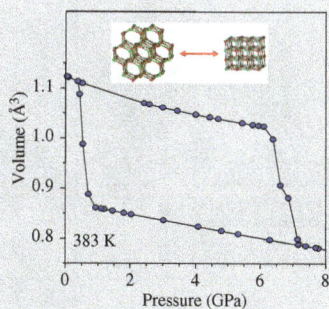

Annu Rev Phys Chem **46**, 595–625 (1995).
Science **276**, 398–401 (1997).

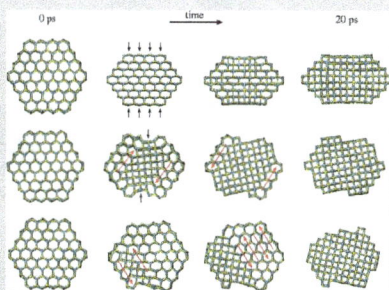

simulations by Madden, Dellago, and Rabani...

Nanocrystals as "Single Structural Domains." A nanocrystal can reversibly change volume without creating defects. This scaling law can be used to design new reversible batteries.

Nanocrystals or nanowires and volume change w/Li intercalation

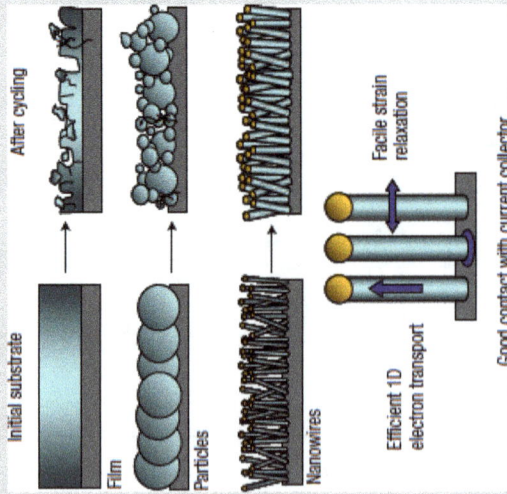

Cui Stanford

Block copolymer electrolytes: dendrite suppression

Balsara Berkeley

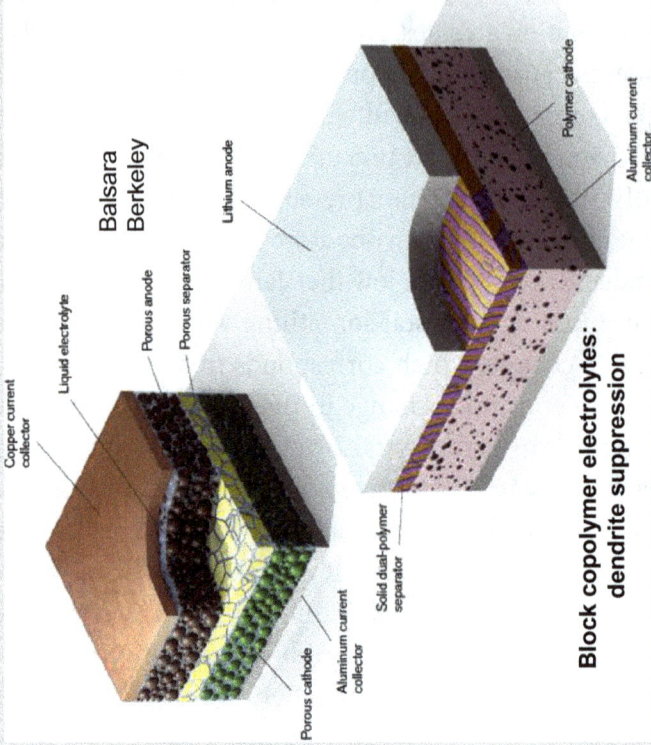

Batteries and control of nanoscale structural domains. Nanotechnology enabled anodes, cathodes, and transport media have the potential to revolutionize battery technologies.

another. If you like you can think of that as a molecule "isomerizing" between two different stable structures. What this means is that the nanocrystal acts as a "single structural domain." This is incredibly important for nanocrystals and it has lots of implications for how we think about them.

Now let's think about that in the context of energy storage. Here's a battery and a former coworker of mine, Yi Cui, who's now at Stanford, one of **Dick Zare's colleagues**, was thinking about this idea of the nanocrystal as a structural domain and he realized, oh my gosh, when I take a battery, what do you do inside a battery? You flow lithium ions from one side to the other, you need to store them some place on the other side of the battery, what you do is you "intercalate" the lithium in an electrode. The lithium will enter into another phase so it will enter into a carbon phase or into a phase of silicon as was described in this particular case by **Yi**. When that happens, the volume of the crystal changes and it might change 20%, it might change even more in some cases, in any case a big volume change.

Well, we know now from our studies of nanocrystals that if we take a very tiny crystal, it will be able to change its volume by that amount without generating any new defects and therefore, I can make a much more reversible structure for storing energy than I could ever have made if I took a big bulk material and tried to do it. So you can see the scaling law is really in action now, it's teaching you how to go and make something that could be more reversible. That's important for cycling a battery many times.

Here's another idea from a colleague of mine at Berkeley, Nitash Balsara. Instead of using inorganic things he's using organic media, but what he has shown is the following problem. You'll be familiar perhaps or I hope some of you will remember. A few years ago, Sony was making the batteries better and better and what they were doing was you take a film and you wrap it into a cylinder. And so you have the source of the lithium and where the lithium goes to, and in between you have some region that the lithium flows through and you

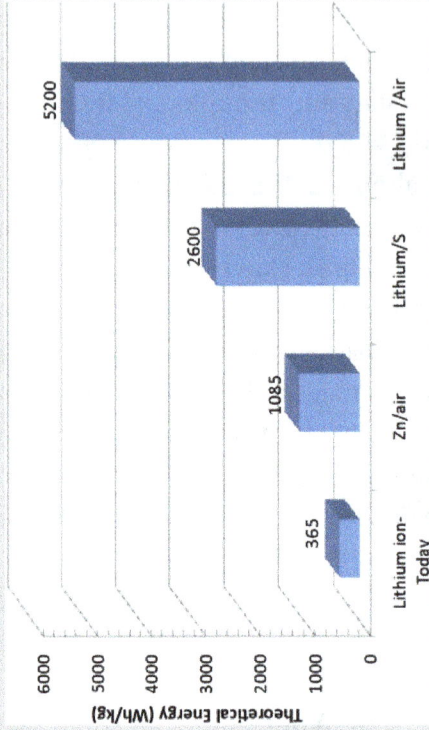

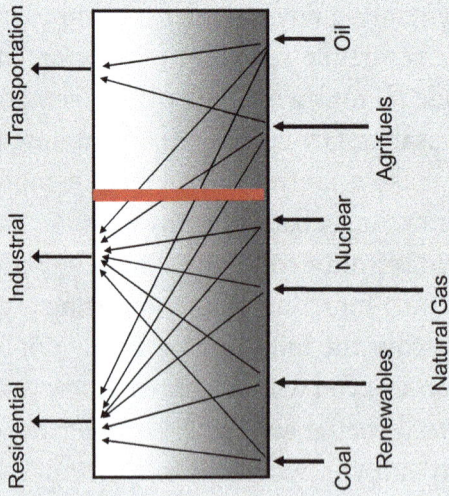

Batteries — gateway to transportation in Carbon Cycle 2.0. today's technology 100–200 W · h/Kg 400 W · h/Kg would give a better driving range.

figure, well I'll just make that thinner and thinner and thinner then the energy density will go up and up and up because when I wrap it all into a cylinder, the same amount of lithium will be transferred but in a smaller volume. So I'll make the separation zone smaller and smaller and smaller. At some point what happened was the battery started to catch on fire. You may remember that. Why was that happening? Why was the battery catching on fire? Well, what would happen is the lithium ions would start to condense into lithium metal on one of the electrodes, a little dendrite. And then the dendrite would sneak through and connect to the other side and short out the battery. And the moment it shorts out, it's all over, current flows through there, a lot of current through a small area, and it gets hot and then it's a combustible material and you get a fire.

Now, Nitash was able to think about this problem, using his knowledge of nanoscience and mechanical properties, in this case of polymer material. And he showed that what he could do is make a mixed material out of two polymers. One would transport ions but it would be mixed together as a block copolymer polymer with another component that is very, very stiff, mechanically quite stiff. So now the ions can move through but the lithium cannot make a dendrite because it can't push its way into the polymer. So what you can see here are some simple ideas which tell you that nanotechnology enabled anodes, cathodes and transport materials can change entirely the way we make batteries. And that's happening today in many areas.

A little bit more about batteries before we go on. Why are batteries so important? And this is now starting to think about how do we decide what problem to work on. Batteries are incredibly important and the reason is the following. If we look at the energy sector, it has inputs like coal, renewables, natural gas, nuclear, biofuels, oil. These are all in energy inputs and then there are energy outputs. What are the energy outputs? Well, a third of energy goes to homes, a third goes to big industry and a third goes to transportation. And it turns out that if you look at this, I can take any of the inputs and feed residential or industrial. I can feed a home with electricity from any one of these sources. But it turns out for transportation, only biofuels and oil can feed transportation. I can't feed it from these other sources presently. And that means it's a huge economic opportunity if you can figure out how to take electricity which would then allow you to have any source feed one-third of the economy of energy. It's a big deal.

Lithium ion it turns out may not be the answer and it may be new chemistries. Today's technology is about 100 to 200 W · h/kg. That may not mean much to you, but it would give you about a third of the driving range of a typical gasoline car. So if we want to have a car that has a decent driving range, we have to get up to hundreds of watt hours per kilogram, which means we probably have to change the chemistry that we use from lithium to something else. So this means the battery area is just full of excitement right now and people are trying to do all kinds of neat things there. Okay, so we've seen now the exercise. We can take a scaling law like melting temperature or what happens when a crystal is very tiny with respect to its volume change and then we can use those to design energy materials. Let's do it with the last two scaling laws, two more scaling laws here. One is going to concern the artificial atom concept, band gap and the charging energy. There are two more concepts there, we're going to see them in action in true energy technologies, okay.

So first the artificial atom concept. This is a crystal of a small semiconductor. And if you look at it you can count the number of atoms across one, two, three, four, a certain countable number across. Now it turns out in quantum mechanics, we know something very important, which is electrons are not just particles, they are waves. So the electron is a wave. And it's an undulating wave inside a big crystal. What happens when the size of crystal becomes comparable to the wavelength of the electron is now the electron will be "confined" inside it and the smaller the box that confines it, the faster the electron will move inside. Just like if you've ever been to one of the science museums where they have those parabolic containers and you can put a ball in it and it will start to roll around and then as it gets in towards the center, it will go faster and faster. That's what happens to an electron when you confine it in a very small box. And its energy levels in a small box become discrete. They separate out and they become "quantum confined" and the energy level spacing scales like one over the box size squared. And that's called the quantum size effect and leads to the "artificial atom" concept. This is a very powerful concept and these materials now are widely used for biological imaging. These small crystals are used because they make colors that don't bleach. Here's an organic dye and it bleaches very quickly. These inorganic nanoparticles which can be introduced inside a cell do not bleach and therefore they're very, very stable.

Now from atoms, I hope you know if we have atoms what's the next thing that you want is molecules, right? So if we have artificial atoms I want to take two atoms and put them together and make a molecule. Then I can build things. And that's what nanoscience wants to do for energy. We want to build things. But at first when you start

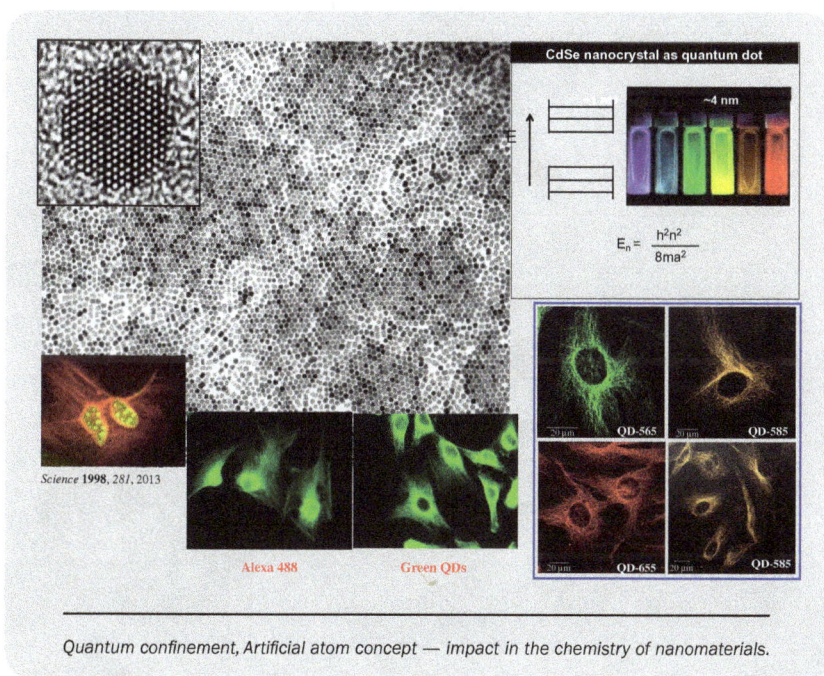

Quantum confinement, Artificial atom concept — impact in the chemistry of nanomaterials.

you have to have a mechanism for building. So for us, in my group, we have spent many years learning how to go from artificial atoms to artificial molecules by developing chemical systems, reactions, which produce reliably certain types of shapes like hollow particles or striped particles or branched ones, or all kinds of little funny shapes like that. And as we learn to form these into a system then it turns out that becomes like molecular chemistry. In molecular chemistry, we can start with a set of molecules and with well-defined rules, hopefully, we can add one to another and build up, for example, a drug molecule. In that same way what nanoscience would like to do is to take artificial atoms and learn how to put them together into intricate arrangements which allow the energy flow to be controlled in a very specific way on a very small length scale. And that's by making an artificial molecule. So that's one of the big goals of nanoscience.

Now Rutherford, the very famous physicist who discovered nuclei, okay, Rutherford once made a very famous quip about science,

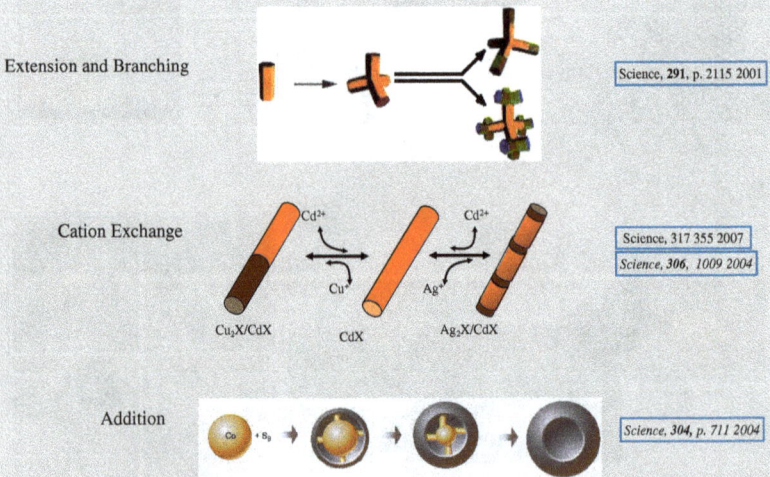

Elementary chemical and physical transformations of nanocrystals: from artificial atoms to artificial molecules

Extension and Branching

Science, **291**, p. 2115 2001

Cation Exchange

Cd^{2+} Cd^{2+}
Cu^+ Ag^+

Cu_2X/CdX CdX Ag_2X/CdX

Science, 317 355 2007
Science, **306**, 1009 2004

Addition

Co + S₈

Science, **304**, p. 711 2004

These operations can be performed sequentially in "any" order
Solid state synthesis now can achieve the complexity of molecular synthesis

when he said there are only two branches of science: physics and stamp collecting. And I don't know whether he felt good about the stamp collecting or not, but it turns out I'm a stamp collector and so are all of my friends. And this is our stamp collection and these are the nanoparticles that we learned how to make. We made branched ones, hollow ones, one completely inside another like Russian dolls, ones that look like P orbitals and F orbitals and stripes and so on. And what I wanted to say to you about this is that I think Rutherford was exactly right in the following sense that when a field of science starts, the first thing that it has to do before you can systematize it is, you have to scan and just see what is possible, what happens, and that's called stamp collecting. So this was a fit. Then the second phase is now you know approximately what the land looks like. Now you can start to perform a deep analysis and understand the principles that underlie it. So that's why his saying makes a lot of sense. And we've gone through the stamp collecting phase, so now we can be in the science phase of building things on purpose because we know how to make all of these structures.

So what would be one we might want to make? Well, here would be a challenge and I said there's a deep connection to biology. I mentioned before that nanoparticles can be used to image biological systems. For example, in the imaging of biological systems today if somebody that you know has a biopsy for a tumor, they take a piece of tissue out to see is the tumor benign or malignant. A pathologist will stain it, in order to image and see what it is, and it is very likely that they will use a nanoparticle to do that imaging in order to see what that is. So that is something valuable, imaging, that nanoscience offers to biology. It goes the other way, too. Here's a very famous biological system, the photosynthetic membrane. And it is an incredibly intricate device which takes photons and creates new molecules. And people cannot do something that rivals that. It's very, very difficult for us. So it's interesting to try to dissect it and see if we could imitate nature, and make an artificial material for photosynthesis. One reason is because then we could make a complete carbon cycle for energy, but

A nanocrystal "stamp collection" ...

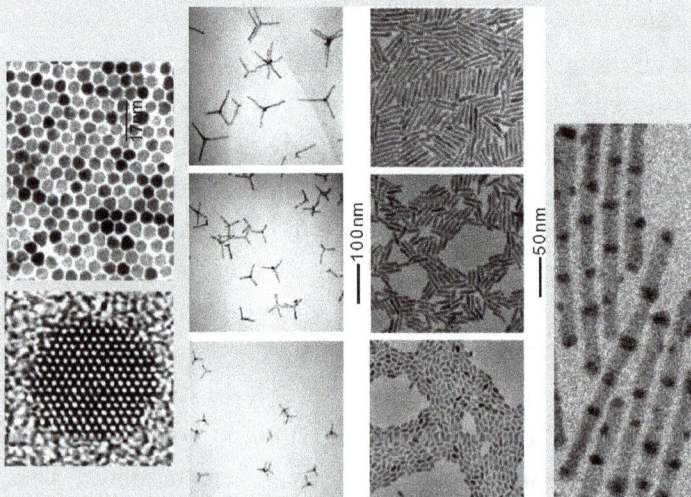

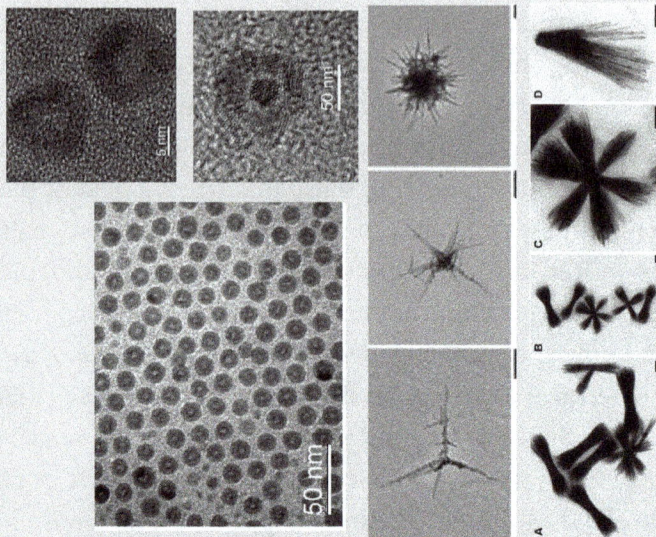

Symmetry, topology, connectivity, directionality...

also because it's a deep intellectual challenge, can we do something as complex as that. And just look at all these arrows, I don't even know what many of them are either. It's such a beautiful complex system. It's got many, many characteristics to it. Could we make an artificial system that has some of those characteristics completely from scratch? Okay, we have to find out.

So let me say some characteristics about what you would like to have happened in an artificial photosynthesis system. You want to absorb light, separate a charge, you will make an electron and a "hole" (the absence of an electron) and then you want to use these to drive a reduction and an oxidation reaction. And in the natural system that's that's driven across a membrane.

And here are some things that we could think about. Now I want to try to make an argument that says, if it's optimized, it probably will be made out of nanoscale components. Here's my argument.

Artificial Photosynthesis and Nanoscience

PSII   Cyt b,f   PSI   ATP synthase

•High density of reactants (photo-generated charges) leads to more products

•10 k_BT dissipation required to ensure directionality of energy flow

•Catalysis and flow of reactants and products must match solar flux

Imagine that I start with a really big semiconductor that absorbs the light. And now I want to do chemistry with it. The chemistry has to take place in one molecule, and that energy of the photon has to suddenly find itself localized into a chemical transformation in just one bond. Okay, so that means that if the excited electron in the solid is a wave and the electron is very large compared to one bond, I have to localize it now again into that one bond and I already showed you it takes energy to localize an electron. So I'm going to lose energy if I do that. Okay, so that means I'd like to use the smallest object possible to make my artificial photosynthetic system. I'd like it to be one bond, say. Well, what would happen if I did that? Well it would be more likely to fall apart (remember how the nanocrystal quantum dots are more stable than molecular dyes?) So there is an optimum nanoscale size for the light absorber.

It turns out it's very important that the energy flow directionally. If I'm going to do photosynthesis I need to have a direction because I can't have the two sides of the electrodes shorting each other out. The reduction has to be on one side, the oxidation on the other. It turns out I can do an entropy calculation and say it costs me a certain amount of entropy to have the energy go in a specific direction. And that turns out to be around 10 KT. That means, I'm going to lose a certain amount of energy every time a photon gets absorbed no matter what. From thermodynamics I know every time the photon gets absorbed I'm going to lose a certain amount of energy. That says if I have a molecule that's got say 20 atoms in it, I'm going to lose that 10 KT and I have to do that billions and billions of times. The odds are that molecule will fall apart are quite significant. So if I take an infinite solid, it will be very, very stable but I have to pay the energy to localize the electron. If I take a really tiny one, I can absorb the photon just fine, it but I'll probably have an event where I'll break the molecule after a little while. So this is an argument why in the end if you look at the natural photosynthetic system, it's operating with 4 or 5 nm size components. Probably that's what we want to do too.

There's one other thing that we have to be able to do, which is photons just keep on coming from the sun and you have to keep up with them. And by this I mean you may not know it but if you look at the plants outside, when the sun comes up, the photosynthesis starts and it tracks the sun, more sunlight, more photosynthesis, and they track each other up to about 15% of the total noon time solar flux. Beyond that, actually the natural photosynthetic system cannot have the catalysis and the flow of reactants and products keep up. It's too slow. So nature throws out 85% of the sunlight! Maybe we can avoid that problem in our design.

So here's an example of trying, in an early stage, to design a photo-catalytic system. We will use quantum confinement, quantum phenomena of confining electrons, and we will see yet another scaling law pop up, which now has to do with "charging energy." So let's make a system which has a direction to it. We will make a nanoparticle which has in it two catalyst particles, one on the side here and one on the other end. And so here's the one side and there's the end. And we'll build the crystals so that there's a little semiconductor here attached to another one. Remember I told you we can build all kinds of stuff. So there'll be one semiconductor embedded in another and these are organized, these semiconductors such that when a photon is absorbed, holes will stay here in this green zone but electrons will like to be in the yellow zone. And electrons will then want to flow here to this blue area which is where the reduction events could take place. And holes will go to this orange attached catalyst, which will then do the oxidation. So then I will have ensured the directionality of energy flow. And one way that I'll do this is I'll make sure that by quantum confinement that this little zone is so small that the electron doesn't want to be inside it anymore. It really prefers to go outside. Remember if I squeeze on that electron, its kinetic energy is going to go up. If there's another place for it to go, it will jump out and go there. So I built my structure in such a way for this to happen. And here you can see that something like this kind of works if I make this distance larger and larger between the seed and the reduction catalyst. The quantum efficiency for hydrogen production in this case goes up and up and that means that it's working better.

Now I told you I have to keep up with the photon flux and there's some technical things here that I don't want to go into it necessarily in detail, but what we can do in this particular case is measure the light emission from one of these structures. And whenever it emits light that means that the electron and the hole — okay whenever it emits light, it means that the electron in the hole recombined. When it doesn't emit light, it means that the electron and the hole are doing chemistry because the electron went to do a reduction and the hole went to do an oxidation and that's a good thing. Now it turns out if there's no catalyst attached, this semiconductor, if I put in more light, it will just emit more light. But when the catalyst particles are attached, remember there's an oxidation one here, and a reduction one here, I see this very funny non-linear dependence to the light flux. And it turns out if I look at it in some detail what I find out is that at this photon flux, when the photons are coming often enough, I reach the point where on this side the oxidation can no longer keep up. I can't take the holes out of the system as quickly as the photons are generating them. And this zone here is actually a zone where the electron chemistry cannot keep up. I start to build up electrons on here.

And I just want to tell you about this one because it's another one of the scaling laws. I promised you a fourth scaling law and that's the charging energy, so here it goes a real quick qualitative explanation, but very important if you want to think about nanoscience, it shows up all the time. Imagine I have a piece of metal and I stick an extra electron on it. What's happened is, now there's one extra charge and I can ask what is the new charge density? It's increased by one charge divided by the volume of that crystal. If I go to add a second charge, it's repelled by the first one, but if the crystal is quite large they'll hardly feel each other because it's a big volume and they won't feel it, but as the crystal gets smaller and smaller once I've added one charge, if I go to add a second one, the Coulomb repulsion will be

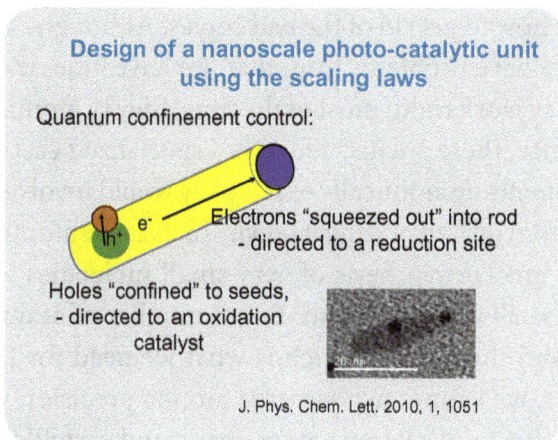

Design of a nanoscale photo-catalytic unit using the scaling laws

Quantum confinement control:

Electrons "squeezed out" into rod
- directed to a reduction site

Holes "confined" to seeds,
- directed to an oxidation catalyst

J. Phys. Chem. Lett. 2010, 1, 1051

great, because they're both confined to the same tiny volume. This is why the charging energy scales like one over the radius, the smaller the radius, the larger the charging energy. These little particles that are doing the catalysis are quite tiny, so small that they can never have more than or two extra electrons on them. In fact they can really only have one. So I can very easily see a saturation event in the photochemistry. If I turn the photon flux up at some point, the catalysis won't keep up anymore, either for the holes or for the electrons. And I can see at exactly what laser power that happens and then I can compare that to the solar flux and try to see, is this going to work or is this not going to work in full sunlight? So that's the exercise that we do to try to make an artificial photo-catalyst work in progress.

So here's the algorithm to build on the nanoscale: design using scaling laws and then measure at the single molecule level. Now why do I say measure at the single molecule level? This is the next very important thing that I want to tell you. This can be a virtuous loop. The reason is because, when we make artificial materials, it's very hard to make two of them exactly the same. Now in nature, the DNA will make a protein and almost every single time they'll be exactly the same, if they're not the natural system has figured out how to repair them or else how to get rid of the bad copies. Although occasionally, you know we have problems from that, we have illnesses from that, but when they work right, most of the time, they're atomically exact. In nanoscience, there are just a few very specialized cases where we can make something atomically exact. They would involve things like picking an atom up and putting it down. That can be atomically exact, or they can involve synthesis of very small molecules which have 10, 20, maybe 30 atoms in them. But as number of atoms becomes large, tens and thousands, which is what we need for these kinds of structures, we cannot achieve true atomic precision yet, but we will. It's possible to. We know nature does it and we will too one day. But how can we get there? How could we get to the point where we could actually make atomically precise things? Well, we'll follow this algorithm and we'll look at each one and see how it's different from the next. But another thing that we need is observation methods which operate at the single molecule level so that we can see what's different from one to the next. So that's a big part of nanoscience is developing new tools.

I'm going to share with you today the tool that we made most recently. This is called a graphene liquid cell and what it is, is just a very simple thing. This is a movie of the growth of nanoparticles and I'll show it to you. Again there's never been a movie like this before. And this is what it's like. We take a piece of graphene, it's a single atomic layer of carbon, and we trap liquid between two of those graphene layers. It turns out graphene in one layer really sticks to the next. So if I have a graphene layer, and I put a drop of liquid on it and I take another graphene and put it on top, what will happen is that the liquid will break up into little pieces, and the graphene will seal around each drop. And now I have a little pocket of liquid that's trapped between the graphene layers and I can do electron microscopy in there and I can go in and look at every nanoparticle as it's forming in the liquid. So here you can see all of its crystal structure and you can observe directly the growth of these particles in solution as it takes place in real time. This is the graphene liquid cell, and it's going to enable us to do all kinds of things in the electron microscope that previously had not really been possible. So it's a way to see.

I'll give you an example of one thing that's seen here. This is a case of two nanoparticles coming together and fusing. And Sir John, I know this is of interest to you. Here is a case where there were certain kinds of defects in the nanoparticles. Here is something that happens very commonly when metal nanoparticles are formed. They come together and they fuse, you might have seen some of those events in the movie of the growth of the nanoparticles. Here we can see two particles that come together, and they fuse. And what we can see is that we can image the actual fusion event and what you can see is I can measure the end to end distance of the combining particles, and the thickness of the neck between them here. And what you can see is as the neck thickens, the end to end distance become smaller, okay. So as the neck thickens the end distance becomes smaller. So I can see that what is happening is atoms are moving around the edges of this particle and coming together and making this crystal. So now I'm actually observing the growth in real time. That's going to enable us to go back later and create much more precise structures than we have previously. So we've seen two, I have a third one if my time is okay. Is my time okay? Five minutes? It's okay.

Okay so let's do capture and sequestration. And this is something I have to tell you about an aspect of the carbon cycle and energy that is very, very important. I mentioned to you earlier 30 gigatons of CO_2 is emitted every year. Now what do you do with a thing like that? 30 gigatons, I mean it's an impossibly large number. You can't really imagine what it means, okay. It turns out you can take the top 100 chemicals produced in the world, the top 100 chemicals, add them all up and they're half a gigaton. So now you know there's a factor of 60 difference between how much energy we use and anything that we make to use, pretty much, okay. This is why the energy problem is so complex. You can't just take the CO_2 that's emitted and say I'm going to make something with it because there's nothing we want on that scale except energy. So if you can make a complete cycle with

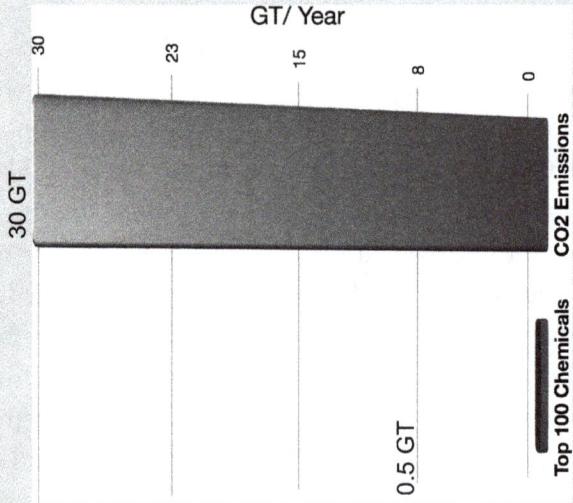

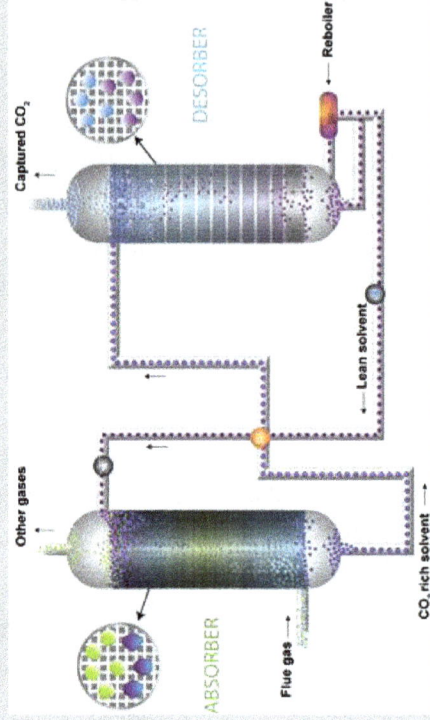

Current amine process energy intensive
~25% of the power of a plant required
~8¢/kWh

Carbon Capture and Sequestration — scale and scope of the problem.

the CO_2, with a photon and make a fuel again, then that matches, it's fine, okay, but absent to that, there's a lot of CO_2 leftover, and we have to put it somewhere. So people are thinking about how to capture the CO_2 and put it somewhere. So let's talk about that.

The current process for capturing CO_2 involves taking the CO_2 and putting it into a liquid water that contains an amine. The amine binds the CO_2. And now you want to release the CO_2. Well, you say why don't you just leave it bound, what you already know why we can't, because there's nothing I could bind it to that we have enough of. So you have to release the CO_2 somewhere away from the atmosphere after you capture it. How do you release the CO_2? Today, you boil the water and that's what we heard about earlier today that, it consumes an enormous amount of energy, therefore it costs too much, about one-third of the energy of the power plant would have to go into releasing the CO_2. It's too much.

So you try to make new materials. So here are another class of nanoporous materials, metal organic frameworks, they're very similar to the ones that was shown by Sir John earlier, but these are materials that combine organic molecules and inorganic atoms in very intricate ways. And in this case my Berkeley colleagues, Omar Yaghi and Jeff Long, have made MOFs which contain carbon dioxide binding amines, but which are solid state compounds, so you never have to boil a liquid. And with these they've achieved now some very promising properties. They can release the CO_2 with one third of the energy that's typically used. So that is a design of a new nanoscale material for energy applications. And in fact you can make many others, but time is not going to allow me to describe them.

So now let's go to the section on energy analysis and risks. So what would you do with all that CO_2 after you capture it, to prevent it from going into the atmosphere and warming the planet. One of the proposals is that you stick it underground. So you take CO_2, you stick it underground and then it stays there. Now why would a geologist think that's a good idea? Well, it turns out our current energy technology is based on taking gases and liquids that have been

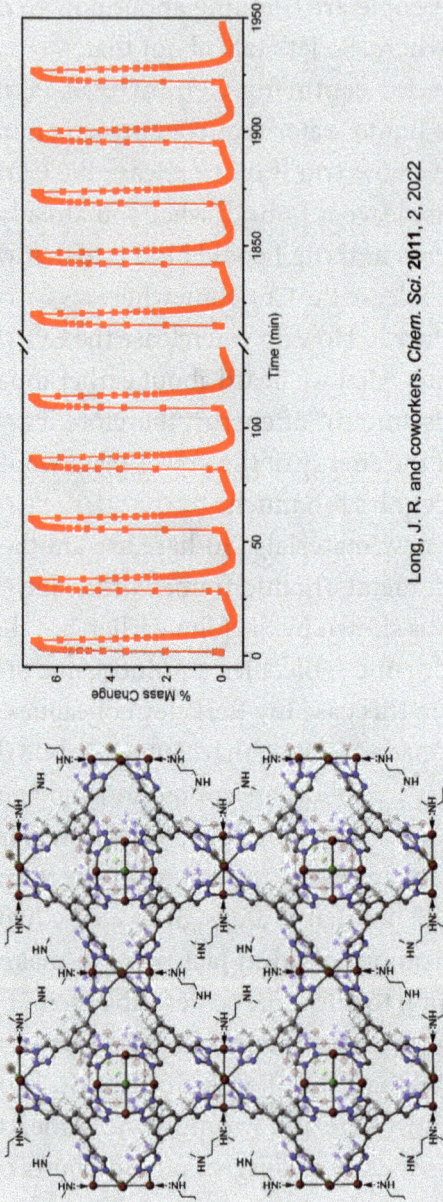

Long, J. R. and coworkers. *Chem. Sci.* **2011**, 2, 2022

Metal Organic Frameworks
Omar Yaghi, Jeff Long

Carbon Capture — New materials. The challenge: develop a material that can capture and release vast quantities of CO_2 at 1/3rd today's cost of ~8¢/KWh — Current technology involves capture using 30% solution of monoethanolamine (MEA) in water with regeneration carried out at 120°C. The working capacity for these solution is also much lower: only ca. 2 wt% CO_2 — We now have a related material that shows cycling at 11 wt% capacity with a similar low regeneration penalty.

CO₂ Saturation at 200 years

MD simulation snapshot of a 7 nm-wide pore (31,000 atoms) at 373 K

H₂O

Quartz

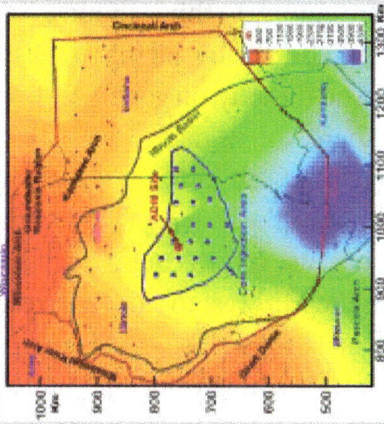

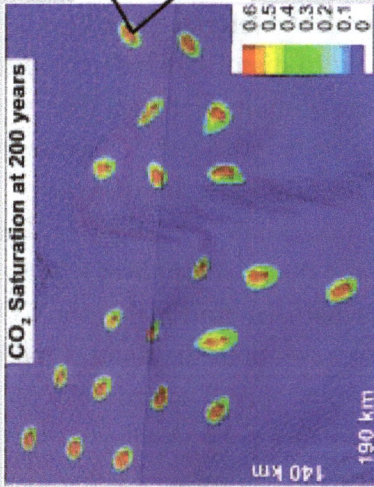

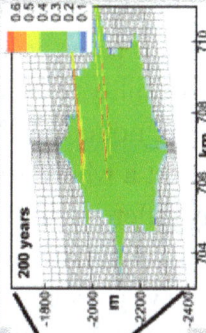

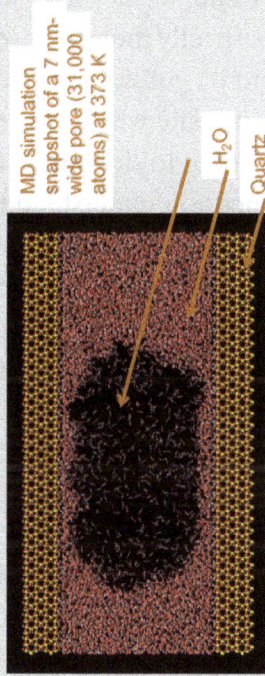

Carbon Sequestration — simulations. Simulated injection 5 Gt CO_2. Over 50 years at 20 sites. No transport between sites, gases move down. Peak pressure rise 30 bar dissipates over time — brine Zhou and Birkholzer GHGST, 2011.

sitting underground for millennia and they're not going anywhere. So if we pull gases and liquids out, we should be able to put others back in and if it's a formation that normally holds gases or liquids, a geologist will tell you it's got an excellent possibility of holding it for a very, very long time. Is that correct? Is that not correct? Well, so at the Berkeley Lab what we do is we now get a whole group of people to think about this from many, many different disciplines. And so for example, they'll try to do simulations. Here's a simulation of CO_2 on a strictly molecular scale. And so you've got some silicate material and it's got water and CO_2 and salt all mixed together and you try to simulate what is the nature of the interface and how is it is the CO_2 going to deposit, is it going to dissolve, what's going to happen. Once you understand the sort of molecular level then you can try to do a simulation at a very large length scale. Here they've taken CO_2 and they stuck it underground in a series of wells in a particular geological formation in Illinois, all in the computer. And they inject a certain amount of CO_2 and they put it in over a period of time at these different sites. And what they find is that the CO_2 injected here, given everything they know about the rocks, what it does is it goes down vertically but it doesn't spread horizontally very much. But what does happen is the pressure goes up a lot.

So then some economists come to help. Now we need economists. We have to have a diverse group, a community to think about it. The economists come and they say okay let's think about all this. We'll do a calculation of the cost of all of this. And they say, oh gosh you know what, you've pushed the CO_2 under there, you've made the pressure high, it's briny water, the water has a lot of salt in it. Actually, why don't you just push the water out, evaporate the water and collect the salt and then take the salt and sell it because it turns out the salt has an economic value. You can use it in the Midwest to salt the streets to melt the snow. And they can calculate the value of that, and it's actually significant. So now you start to see how you could build a system where the different parts talk to each other and one thing you do on one side starts to help with something you're trying to do on the other, but only if you thought about it, because otherwise one thing you do on one side can cause trouble on the other side. And this is why we have to change our thinking to be able to think as a system of energy, and not to think of just one part of the problem. It's very tricky, not at all easy to do.

So in fact now they start to think of all these other ideas, the geologists, and they say well gosh you know, there's another thing that I could do. It turns out I can take air and compress it into a big reservoir and I can store energy that way on a massive scale. And then they say, okay well what's the pressure-volume curve for that? And they say, if I put CO_2 in at pressure, the CO_2 can go supercritical. And when that happens that changes the pressure volume relationship and turns out you can store more energy if you have 20% CO_2 and 80% air then you can't if you had only air. That's very interesting, because now it means that I could take those air energy storage reservoirs and start using them 20% for CO_2 and I could actually have a positive solution again. So you start to see how the community really thinks about interactions in complex ways that are really very, very interesting.

The very last topic here. I was asked to talk about health effects of nanomaterials and I do want to. They're very interesting and they're important, but it turns out we don't know near enough about it. So what can I say about this. Here's an image from my lab from about 10 years ago. This shows, a cell dividing, and it had quantum dots in the "mother" cell. And so the two daughters cells have separated we see that the quantum dots are going into the daughter cells. At the time we were very excited about this because what it meant was that we could take a certain cell type and load it with quantum dots, for example, it could be healthy cells and they could divide and we could have a separate set of cancer cells in the laboratory and they could divide and there wouldn't mix their quantum dots from one to the next and therefore we could track, for example, the cancer cells as they move through a tissue because they'd be the red ones inside a green or something like that. But when you look at a diagram like

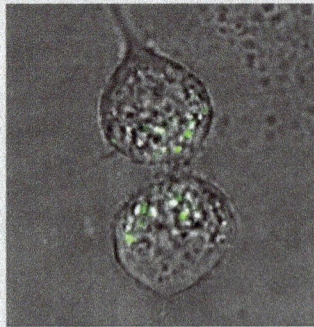

From "Adverse Effects of Engineered Nanomaterials,"
by Fadeel, Pietroiusti, and Shvedova, Academic Press 2012

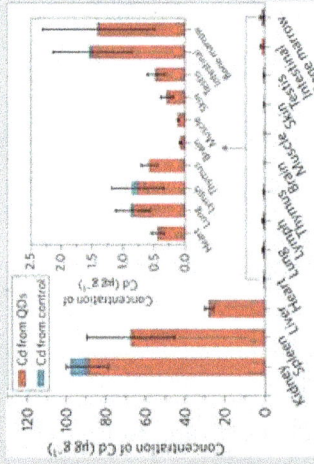

"A pilot study in non-human primates shows no adverse response to intravenous injection of quantum dots"
Nature Nano May 2012

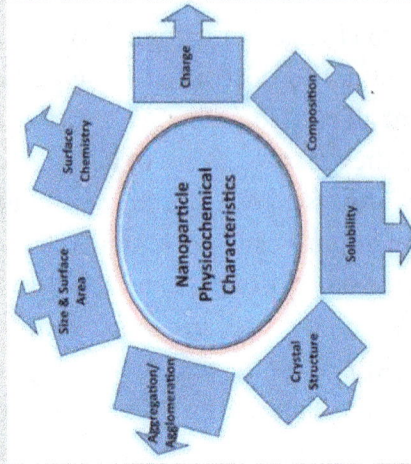

Quantum dots in cell division

this, you start to think, hmm, I wonder if this is all okay or not because there was this inorganic thing sitting around inside these pockets and the nanoparticles just go right in their too very easily and you start to think, well what are the facts of that. And it turns out that people don't completely know. It's a very nascent early topic. Here's a book just came out. This is the first book I know of, that's a real book on this topic, Adverse Effects of Engineered Nanomaterials. And one of the things that it points out which I think everybody in our business knows which is if you have a nanoparticle, it's one thing to say, oh I have a quantum dot, but I can make a quantum dot that has completely different surface chemistry such that it will like to go inside a cell or it will run away. And therefore I have to take into account all of these properties of agglomeration, surface chemistry charge and so on before I could even begin to know where a nanoparticle goes. There are studies which show things like this shows nanoparticles which were injected into some monkeys in this case, and some of the studies have gotten to that point and they follow where the nanoparticles remain as a function of the time. In this particular case a paper that just came out couple of days ago. Pilot study in non-human primates shows no adverse response to intravenous injection of quantum dots. Well, for every paper like this you'll find another one that says something rather different. It's a very chaotic moment. It's a very chaotic moment. The science isn't settled yet.

So that the UK Royal Academy in 2004 issued what I think is the best statement about this to date that hasn't really changed. Treat nanomaterials as if they are hazardous. Seek to remove them from the waste stream. Don't apply free nanomaterials to the environment unless the benefits clearly outweigh the risk. Now what do we mean by free nanomaterials? That would be like particles that are just free to move about, but for example, in a battery they're not going to be free to move about, they're going to be embedded and you can recapture that material and use it later on. So for the vast majority of uses, in fact we can use these materials in a way that I think will be quite reasonable but in the meantime this science has to advance or else we're eventually we're going to get something very, very serious. And they say something very serious here as well, assess potential impact throughout the lifecycle of the nanomaterial. And now I think is a key thing and we certainly need to do that.

Okay, so I'm going to wrap up here. I hope that you have seen the scaling laws, the synthesis and properties of the building blocks, the power of one that means we have to have tools for looking at single particles and their trajectories and what they do, the very deep connection between artificial nanomaterials and biology both benefiting both biology and nanoscience. And right now we're in this era of we just learned how to make stuff, so now we're practicing making it and we'll just have to see what comes out of that, but here's a very, very famous phrase from another person, I said Democritus was the father of nanoscience okay, a more recent candidate would be Richard Feynman, a famous physicist, whose lecture on nanoscience really assured in the field in some sense and a very famous phrase of his, "what I cannot create I do not understand". That's a kind of model for nanoscience and it exemplifies what we try to do in this field and how it works. Thank you.

Enzymes by Evolution: Bringing New Chemistry to Life*

Frances H. Arnold[1]

Not satisfied with nature's vast enzyme repertoire, we want to create new ones and expand the space of genetically encoded enzyme functions. We use the most powerful biological design process, evolution, to optimize existing enzymes and invent new ones, thereby circumventing our profound ignorance of how sequence encodes function. Mimicking nature's evolutionary tricks and using a little chemical intuition, we can generate whole new enzyme families that catalyze important reactions, including ones not known in biology. These new capabilities increase the scope of molecules and materials we can build using biology.

Keywords: Enzyme Evolution; Directed Evolution; Carbene Transfer; Organosilicon; Biocatalysis.

---

* This chapter was based on a lecture delivered by professor Arnold at the Molecular Frontiers Symposium in the Royal Swedish Academy of Sciences in May 2017.
[1] Nobel Laureate in Chemistry (2018). Division of Chemistry and Chemical Engineering, California Institute of Technology, Pasadena, CA 91125, USA.

About Frances H. Arnold

Frances Hamilton Arnold is the Linus Pauling Professor of Chemical Engineering, Bioengineering and Biochemistry at the California Institute of Technology, where she pioneered methods of directed protein evolution used to make better proteins for applications in alternative energy, chemicals, and medicine. She earned a B.S. in Mechanical and Aerospace Engineering from Princeton University and her Ph.D. in Chemical Engineering from UC Berkeley. She carried out postdoctoral research in biophysical chemistry at UC Berkeley before coming to Caltech in 1986.

Arnold's work has been recognized by many awards, including the Nobel Prize in Chemistry in 2018, the Millennium Technology Prize in 2016, the U.S. National Academy of Sciences' Sackler Prize in Convergence Research in 2017, the Charles Stark Draper Prize of the National Academy of Engineering in 2011, and the U.S. National Medal of Technology and Innovation, which she received in 2013. She has also been elected to membership in the U.S. National Academies of Science, Medicine, and Engineering, the American Academy of Arts and Sciences, the American Philosophical Society, and the National Inventors Hall of Fame.

Arnold chairs the Advisory Panel of the David and Lucile Packard Foundation Fellowships in Science and Engineering and serves as a Trustee of the Gordon Research Conferences. She is also a Director of Illumina. Co-inventor on 58 issued patents and active in technology transfer, Arnold co-founded Gevo, Inc. in 2005 to make fuels and chemicals from renewable resources and Provivi, Inc. in 2014 to develop non-toxic modes of agricultural pest control.

I want to explore biology beyond the natural world, using evolution as a tool to create things that nature has not made. Thus this talk is about bringing new chemistry to life, making enzymes that catalyze reactions we have not (yet) found in nature. You might think there is already enough interesting chemistry out there in the biological world to keep us busy and satisfied. I will argue, however, that human chemists have been pretty clever over the last one hundred years, discovering catalysts and making useful molecules, and it would be marvelous to be able to genetically encode some of their inventions, or, even better, go beyond what both humans and nature have been able to.

I started my career as an engineer: I have a degree in mechanical and aerospace engineering, and I worked at the Solar Energy Research Institute, newly formed when Jimmy Carter was president. In the late 1970s, my country actually had a national goal of 20 percent renewable energy by the year 2000. Sadly, we didn't stick to that. When the election of Ronald Reagan forced me to rethink my plans in solar energy, I went to graduate school for a PhD in chemical engineering. My timing was perfect: I was at U.C. Berkeley at the beginning of the DNA revolution. Imagine how exciting it was to be a student when brand-new companies like Genentech, Chiron, and Amgen were starting up, all built on the idea that we could manipulate the biological world at the level of DNA and rewrite the code of life to solve human problems. I was inspired by this remarkable technology to dream of new ways to make the materials, fuels, and chemicals that we need in our daily lives.

The more I learned about enzymes, the more impressed I was by how these amazing biomolecules take materials and energy from the environment and convert them into self-repairing, self-reproducing, responsive, adaptive...life. Enzymes are the most beautiful of all the molecules and are so well designed by evolution to do chemistry better than any human. When it comes to catalysis, there's nothing that can beat an enzyme.

```
AATAGCCGTTATTTCCGGATGTGCATA
GCTGATTTGACCCATCCGGTACACCA
ATDACAAATCCCGATTTGATCGTGTGC
GCGACATGTCTTCCGGCGACACATGT
GTCTCTCACTCCGAGAGATCGGTTAG
AGTCTCGGTTAACCACACGTCCCGGA
TATATTTAATTGGCCGGAGAGTCTCCC
GCGCGACATAAGGAGTCCTCGTTTCG
AGATACGTACGGCATGGTGACACCAG
TTGCCCTCTGATTCCCGGAGCCTCTTT
GAAAACGTCGAGTCGAATCGAAGTTC
GAACCCCGGATCGGGTCCACCAACTT
AGAGATGTGTGTGCGCTGACTCAGTC
ATTTTGACCCATCCGGTACACCAATGG
GTCGGACAAATCCCGATTTGATCGTGT
GCGCGACATGTCTTCCGGCGACACAT
GTGTCTCTCACTCCGAGAGATCGGTT
AGAGTCTCGGTTAACCACACGTCCCG
GATATATTTAATTGGCCGGAGAGTCTC
CAATAGCCGTTATTTCCGGATGTGCAT
AGCTGATTTGACCCATCCGGTACACC
AATGGGTCGGACAAATCCCGATTTGA
TCGTGTGCGCGACATGTCTTCCGGCG
ACACATGTGTCTCTCACTCCGAGAGA
TCGGTTAGAGTCTCGGTTAACCACAC
```

If we could rewrite the code of life, or even just the code of enzymes, then we could look at an organism, say a yeast or a bacterium, as a chemical factory that uses renewable resources and converts those into valuable products, all programmed in the DNA. There's the ultimate, programmable, green chemical synthesis machine! This is a vision that has been around for 30 years or more. It's been hard to implement, though, because it's very hard to compete with pumping oil out of the ground and using that as a feedstock for fuels and chemicals. But if we want sustainable ways of making these critical substances, we have to be able to compete. That means we have to compose new DNA so well that we can subvert three-plus billion years of evolution that says 'make more microbes' and instead have a microbe make the chemicals that we want. That's a very difficult DNA design problem.

And of course my colleagues, who are pretty good chemists, say, "Oh, Frances, you know biology is great, has lots of interesting

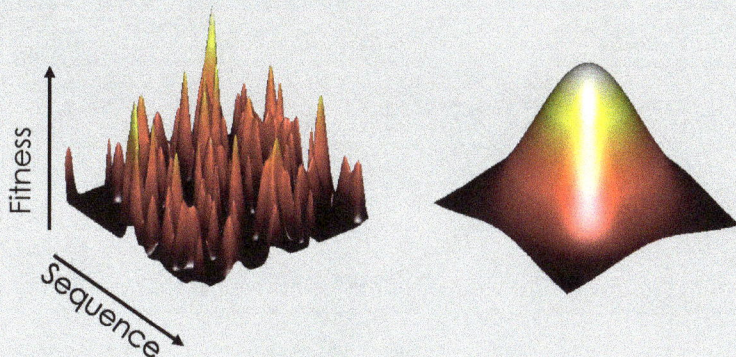

Directed evolution is a molecular optimization process on a multi-dimensional fitness landscape, where fitness is performance, defined by the user.

chemistry, but it can't do what I can do." And they are right, because the chemical industry is based on chemistry, not biology. Ours is a synthetic world, full of products made by synthetic chemistry, containing molecules and chemical bonds you don't find in the biological world. Biology can do some of the things that human chemists have come up with, but many others have no biological precedent.

So here's what I want to do about it: I want to be able to genetically encode chemistry that has been invented by human beings. And not only that, I want to go beyond where humans have been able to go chemically and do it in a sustainable fashion. This hard. The enzyme design problem — that is, how you design a good enzyme catalyst that does not already exist in nature — has been around for as long as we have been able to manipulate DNA. Progress in creating new enzymes has been stymied by two major hurdles on the path between a DNA sequence and what the resulting protein does. First, how does the sequence encode a folded protein? The field has been making good headway on this 'protein folding' problem, but in fact that's not the problem anybody cares about in my business, because we are all focused on the more important

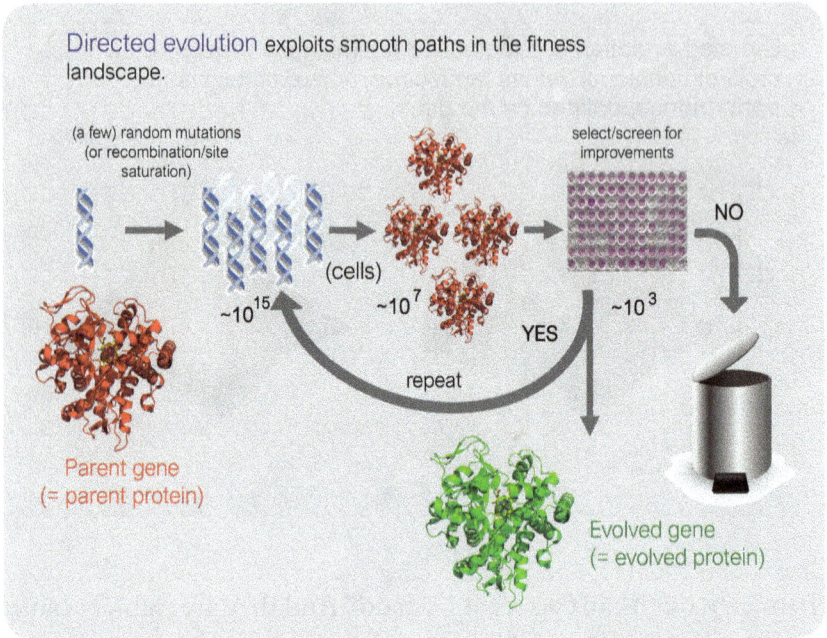

Directed evolution exploits smooth paths in the fitness landscape.

(a few) random mutations (or recombination/site saturation)

select/screen for improvements

NO

(cells)

~10^{15}

~10^{7}

~10^{3}

YES

repeat

Parent gene (= parent protein)

Evolved gene (= evolved protein)

issue of how a sequence encodes function. And that function needs to be good enough to compete with pumping oil out of the ground or at least your favorite synthetic chemist. That's a very high bar. Thirty years ago people said that they would have this problem solved in 5 years, and it still hasn't been solved today. Thus when I started out as a protein engineer, I had to take a different tack, not the one all the protein engineers were using, which was to get a crystal structure of your protein, use your big brain to identify where you're going to make changes, make those changes, and take two months to find out it doesn't work.

The technology of molecular biology has progressed tremendously in the past 30 years. You can now synthesize in the test tube essentially any DNA you want. You can type your sequence into your computer, e-mail it to your favorite supplier, and in a few days or weeks you will get in the mail the actual DNA. It's amazing. So now we can read DNA, we can write DNA, and we can edit DNA. But no one knows how to compose DNA. No designer did

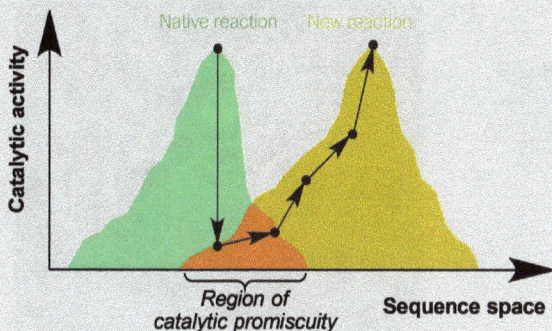

Novelty is already there. Enzymes are 'promiscuous' and can catalyze new reactions (albeit at a low level).

Access to new functions relies on some (even very small) overlap that can be drawn out and optimized during evolution.

Native reaction   New reaction

Catalytic activity

Region of catalytic promiscuity   **Sequence space**

it: three billion years of evolution gave us these wonderful catalysts that you can scrape off the bottom of your shoe and that are better than anything a human can design.

In fact, we would not be here unless evolution did a very good job at making all those wonderful proteins that we can use, enjoy, and even hope to make better. If you look at the sequences of homologous enzymes that exist today, you can build their family trees and see how they came from a common ancestor through a process of mutation and selection. This simple algorithm of evolution created all of these proteins so beautifully tuned for their biological roles. Nature figured it out. So I asked, why not use this design algorithm to build new enzymes?

This is not a new idea, at least in part. I want to use evolution for forward engineering, using all the new tools of molecular biology. But humans have been manipulating DNA, rewriting the code of life, for thousands of years by choosing who goes on to parent the next generation. We've been making corn, carrier pigeons, lab rats, race horses, you name it, we have been manipulating the biological world at the level of DNA to solve human problems, such as how

Can a heme protein do this? Does it evolve?

Form the Fe-carbenoid intermediate

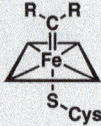

Transfer the carbene to a second substrate

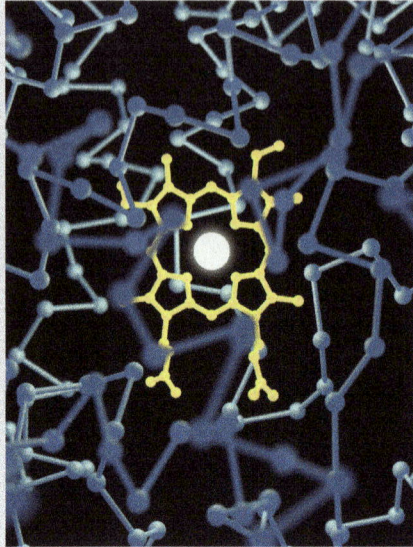

The heme protein is a self-assembling, DNA-encoded, chiral metal complex whose structure and electronic properties can be tuned by mutation.

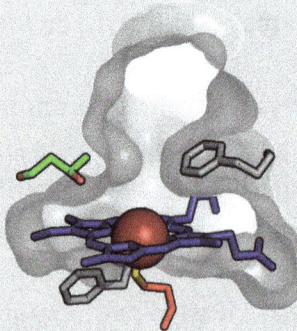

do you feed a population and have time left over to do science? We have been breeding organisms that help us or that merely give us pleasure, and these are not remotely 'natural.' This poodle would not be here without the intervention of human beings. If it were to be released in the wild, at least in my city, it would be eaten by the coyotes. We have been using this process of breeding and artificial selection effectively and to our benefit for quite a while, all without understanding the details of how sequence codes function.

Now, of course, this is limited to evolution and breeding of the biological world, where the sequence diversity is limited. Worms mate with worms, monkeys with monkeys, but you don't cross the two. Maybe you can increase your random mutations by taking more airplane flights or smoking a few more cigarettes, but you can't really control the mutation load. But remember, I said you can now create any DNA sequence you want. With 'molecular breeding' you can choose 3 parents or 33 parents, and you can dial in any level of mutation. You become the breeder of molecules and have much more control, for good or for bad, over this process. The problem is that we have all this control, but we don't know the rules. You might think it's a hard problem, because for even a single protein a few hundred amino acids long and 20 amino acids to choose from

Selective cyclopropanation at scale: replace precious metal catalyst and multi-step synthesis with a single step that uses Fe-based catalyst

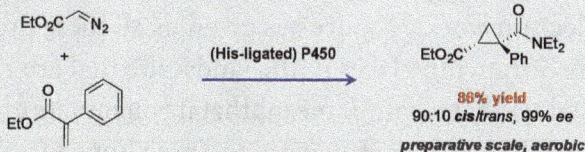

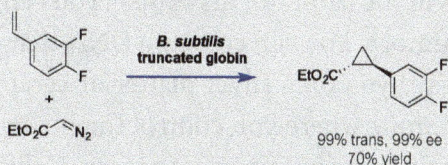

(His-ligated) P450

86% yield
90:10 cis/trans, 99% ee
preparative scale, aerobic

B. subtilis
truncated globin

99% trans, 99% ee
70% yield

BRILINTA.
ticagrelor tablets
(Key intermediate)

Wang, Z., H. Renata et al., Angew. Chem. Int. Ed. 53, 6810 (2014)
Hernandez, K. et al., ACS Catalysis 6, 7810 (2016)

provivi

at each positions, that's a really big space of possible sequences. It's bigger than the number of particles in the universe. It's bigger than the national debt of the United States. It's a really big number, and it's mostly empty: most of those sequences don't encode anything that solves a human problem, and they certainly don't compete with pumping oil out of the ground. How do you search a space of enzymes that's bigger than you can even begin to comprehend and mostly empty? It might look like an insurmountable optimization problem until you think a little more about the nature of the fitness landscape. If the landscape is smooth, you can take a random uphill walk to the top, but if it's very rugged, every time you make a mutation you fall off into the abyss. John Maynard Smith thought about this problem back in 1970 in my favorite, two-paged *Nature* paper. And this was before all those sequences were available to help us to see how protein have evolved. He said the landscape has to be smooth in at least some dimensions, otherwise we wouldn't be here, as we are the products of evolution. Some mutations are 'lethal', but many are not. So evolving proteins wander around in this fitness landscape, accumulating mutations and sometimes changing function. Thus the fitness landscape, even though it's mostly empty, is locally smooth in enough dimensions that evolution by gradual change and selection works. Nature has given us all these proteins that are already occupying places in the landscape that are rich in function. What does that mean? It means that if you mutate existing proteins, you can change their sequences and even hope to improve them for new tasks.

Implementing this in the laboratory is simple: you can even do it in a high school laboratory. You can copy the DNA that encodes a protein that nature gave you on a silver platter or scraped from your shoe, under conditions where you control the incorporation of errors. It's easy to make a lot of imperfect copies. Cells then do the hard work: when you put that DNA into microbes, they will start reading it and make proteins that have mutations in them. And you, the breeder of molecules, then have to decide who goes on

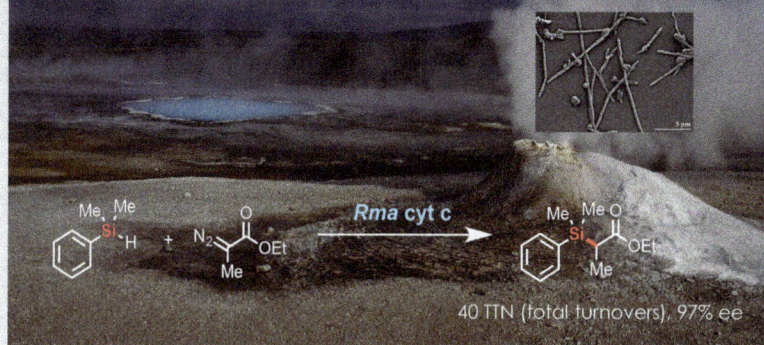

Rhodothermus marinus cytochrome c makes C-Si bonds

Gram-negative, thermohalophilic bacterium from hot springs in Iceland

124 amino acids, denaturation (melting) temperature T_m = 106 °C

Native function is electron transfer

Rma cyt c

40 TTN (total turnovers), 97% ee

to parent the next generation, using good old-fashioned analytical chemistry to measure how the proteins behave and choose one(s) that is/are better than what you had before. You really don't have to look at huge numbers of variants to find improvements, if you've asked the right kind of question. You have to be able to measure the small improvements expected from small sequence changes (1 mutation), and then you iterate on this process in an uphill walk to accumulate beneficial mutations and optimize for a new, evolved feature.

I'm going to quickly summarize the work that lots of people have done in this field before I tell you what is new. Proteins adapt very quickly by this process. It's amazing how well it works, it's the stupidest optimization strategy you can think of — a random uphill walk — but it has generated thousands of useful new proteins that are tuned for what humans would like them to do. The scary fact is that when you try to reverse-engineer the products of this evolution, you find mutations, for example, that are 30 angstroms away from the active side of your enzyme. No one can even explain how those mutations improve catalytic activity, but they do. This just goes to show that we are far away from being able to predict them. Proteins are complicated, but evolution takes care of all of those details. And evolution is great for optimization.

Now, I promised you that I would address the evolution of new chemistry, not just optimization of what nature has already given you. Creating a new enzyme could appear to be a very a hard problem, where you have to perfectly position multiple functional groups simultaneously to catalyze a new reaction. But it's not such a difficult problem if you think about how nature does it. Nature is making new enzymes all the time. I'll just give you two quick examples. We dump potent herbicides like atrazine into the natural environment. We think they are non-biodegradable and then Boom! Some organism has found out that if it removes this chlorine atom, it has a rich new nitrogen source that gives it a selective advantage. Then the organism moves all over the world, and atrazine now bio-degrades. All because a soil microbe discovered a new de-chlorinase activity. And it found that activity in an existing, catalytically promiscuous enzyme, whose biological function was something else altogether. Here's another one that you might have seen in recent headlines: organisms are learning how to eat plastics that we have been dumping into the environment only for 50–60 years. The organisms and their enzymes are adapting rapidly to take advantage of this new opportunity, although perhaps not rapidly enough.

Also think about antibiotic resistance. Nature is solving all sorts of interesting problems in real time. And how does nature do it? Nature solves problems because the capabilities to do so are already there: novelty is already there. Like enzymes, each one of us has multiple capabilities. Someone may be selected for their ability to work long hours in a laboratory, but they can also play the piano or do dishes if necessary. Enzymes, too, have multiple capabilities that are not being used at a given time. But when a new opportunity arises for it to shine, the enzyme is poised to do it. This

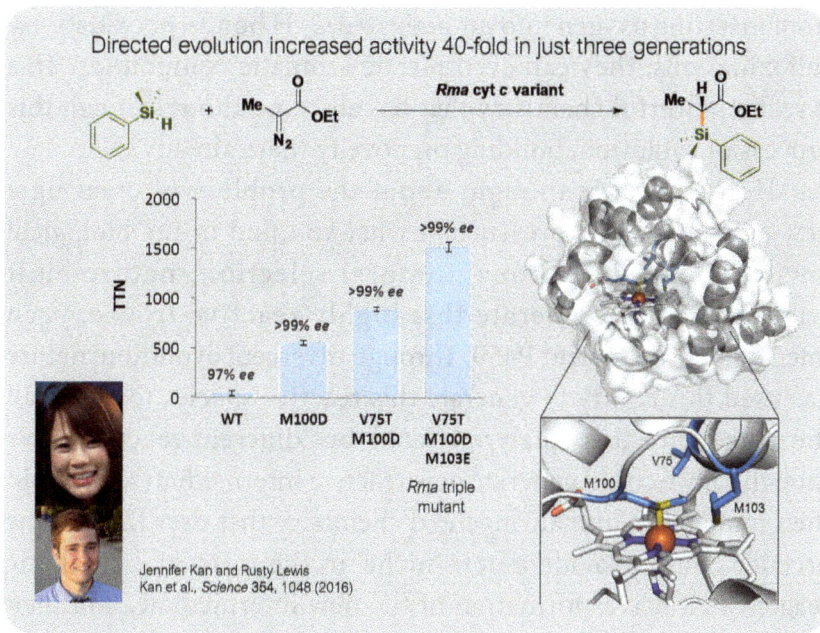

Directed evolution increased activity 40-fold in just three generations

Jennifer Kan and Rusty Lewis
Kan et al., *Science* 354, 1048 (2016)

new activity becomes beneficial and can be optimized through the random uphill walk process that I just described.

People talk about the Internet of Things. Well, biology is the ultimate Internet of Living Things, a gazillion organisms out there working 24-7, trading genes and crowd-sourcing to solve new problems. It makes sense to use that process to move into the future. But how do I breed forward novelty? How do I use this to create a new enzyme?

A good place to start is with a family of enzymes that has already done it many times. Here is my favorite family, the cytochrome P450s. You have ~50 of these in your own bodies; they are your first line of defense against various noxious things you ingest. They detoxify chemicals by sticking oxygen in different places and making them easier to excrete or degrade. This enzyme family has gone through mutation and selection not only for optimization of function but also to take on new and different functions. P450s catalyze a whole slew of impressive reactions, from inserting oxygen into un-activated C–H bonds, epoxidations, sulfoxidations; they can even nitrate aromatic compounds. This is really powerful chemistry that has all come about through this process of evolution, building on novelty that's already there.

Here's how we thought about the problem of creating a new enzyme. In the presence of what you find in the biological environment and through natural selection, nature built a mechanism to generate this highly reactive iron-oxygen intermediate in ancient P450. Through divergent evolution, nature has used this ability to generate this reactive species to create all those enzymes that catalyze related, but different reactions. We thought, if we could generate new reactive intermediates, we could then implement human-invented chemistry that uses those. I the breeder could provide a new 'niche' in the form of a synthetic reagent that drives formation of the new intermediate, and then use directed evolution to optimize any new activity that appears. In this way, I should be able to make enzymes that catalyze human-invented reactions that are not known in the biological world.

Here's an example. Chemists have figured out that this iron-porphyrin-carbene intermediate can be used to convert an alkene to a cyclopropane. Or you can insert the carbene into a nitrogen–hydrogen bond and make a new carbon–nitrogen bond. That is interesting chemistry that nature doesn't do, but has been achieved with small-molecule catalysts. We asked: can a heme protein form the iron-carbene intermediate? (There was some evidence indicating that it could: carbenes can be good enzyme inhibitors.) But then instead of transferring the carbene to itself or catalyzing dimerization, could the enzyme transfer it to a second substrate, and could that catalyst evolve? The carbene precursors we provide aren't prevalent in the biological world, so nature perhaps never had a chance to do it. But, in fact, if you add styrene and this carbene precursor to various heme-containing proteins, you'll find that they all have a little bit of activity. Even free heme will do it in water. So will cytochrome c and myoglobin. And cytochrome P450 has a

One enzyme: 20 example products, most are enantiopure

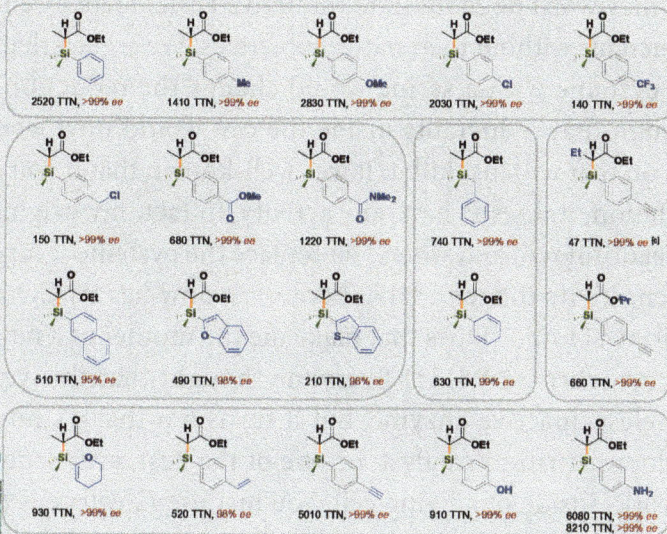

miniscule bit of activity. But, remember, for me, that's the starting point for evolution, because the novelty is already there! If I see a little bit of activity, I'm happy, because I can improve it with directed evolution. And, in fact, if you substitute one of the P450's amino acids with another one, activity goes up sixty-fold. Styrene and EDA can make four possible cyclopropane stereoisomers, but this enzyme only makes one. Other mutations allow you to make other products, such as a *cis* isomer, which is hard to make using small-molecule catalysts.

Now, I'll bring it to life. This new enzyme is encoded in a DNA sequence and should be able to work inside of a cell. Some of you might note that we have dithionite here to reduce the enzyme to its active form, and that's not remotely compatible with a living cell. We try not to contaminate our beautiful evolution experiments with thought, but every once in a while it's beneficial to do just that. Every P450 on the planet has a cysteine proximal ligand. But these enzymes are so beautifully gated that inside the cell they won't be reduced to the active Fe(II) form by endogenous reductants without their native substrates. We argued that replacing the cysteine with a serine would change the reduction potential enough that reductants inside the cell would now work. People say, oh that will just kill it. It was well-known that if you make that mutation, it doesn't have any activity. In fact, my original protein is beautiful red, and when you replace the cysteine it turns a sickly green. It's no longer a P450 because it now has its CO-difference peak at 411 nm. (Does that make me the mother of a new enzyme, the cytochrome P411?) Everyone thought that this version was a useless, inactive enzyme, but it turns out that it's now the best cyclopropanation catalyst, or one of the best, ever reported. And it works inside of a living cell. You just use *E. coli* cells to catalyze the reaction. With seventy thousand turnovers, and the *cis* product with beautiful enantioselectivity.

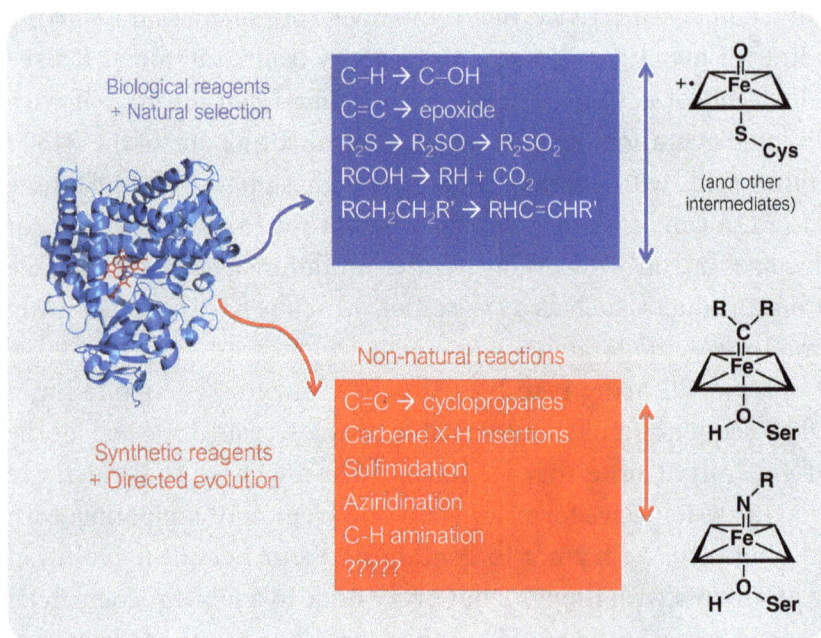

It turns out that the P411 is a very special enzyme. It's hard to convey this if you don't have love for some of these molecules, but I'll try to explain. This new enzyme, which everybody thought would be dead, is not dead. It's just waiting for you to ask what it can do! For one thing, the P411 can also transfer nitrenes. We tried a lot of enzymes for nitrene transfer, but only found the P411 with this non-natural activity. Two post-docs in my group have shown that you can make a P411 enzyme that catalyzes direct intramolecular C–H amination, and the enzyme can react selectively with one of these C–H bonds versus another. These two C–H bonds are next to one another but differ in their bond strengths by more than 5 kcal/mol. With small-molecule catalysts, you always go to the weak one, and in fact, our first enzyme also did that. But what's really wonderful is that you can use directed evolution to accumulate mutations around the active site that promote reaction at the stronger bond, right next door. That's really impressive, and that's what gets us excited, because we can talk about new chemistry that's really selective.

The next challenge, intermolecular C–H amination, is really hard, because the enzyme has to generate the nitrene and transfer it to a second substrate before the nitrene gets reduced. This enzyme has a reductase domain that helps with the initial reduction to the active state, but then also happily reduces the nitrene. So, most of these proteins burn money, so to speak, by reducing the azide. But if you find a little bit of intermolecular amination activity, you can evolve the enzyme and reshape the active site so that this now becomes preferred over the money-burning route. We first demonstrated that you can get intermolecular sulfimidation and aziridination to form these three-membered, nitrogen-containing rings. And just recently we achieved direct (benzylic) C–H amination. None of these reactions is known in the biological world, but they're useful synthetic reactions, and they take place inside or outside of cells.

I hope you can now look at a protein with a whole new set of eyes. We see a heme protein as a self-assembling, fully

DNA-encoded catalyst whose electronic and steric properties can be tuned by evolution. It's an iron complex that can be modified simply by mutation of the DNA to make it capable of all sorts of interesting chemistry. For example, we tuned a P450 to make, in a single step, the *cis*-cyclopropane core of this FDA-approved drug, Fetzima. You get beautiful selectivity for a single stereoisomer out of the four possible. Another example of great selectivity is the ability to make the key *trans*-cyclopropane intermediate to another drug, ticagrelor. A single biocatalytic step at room temperature in aqueous solution replaces five or six chemical steps, often involving a precious metal, and provides an environmentally-friendly way to make these drugs.

I'm going to end with one last example, a recent story that you might have seen as it received some attention in the press. Consider silicon. Maybe you've wondered why life does not incorporate silicon into its organic compounds. We find no carbon-silicon bonds in biological molecules, even though silicon is the second most abundant element on the planet's crust. (The most abundant, oxygen, is tied up with a lot of silicon in rocks.) As far as we know, nature has not made enzymes that forge carbon–silicon bonds, or has she? Chemists have invented an array of organosilicon compounds to make sealants, caulks, lubricants, cosmetics and chiral molecules that show up in catalysts, imaging agents, and even LEDs. The history of small-molecule transition metal catalysis for enantioselective carbene insertion into Si–H bonds is a short story. There are a few catalysts reported to catalyze carbene insertion, but notice the metals: iridium, copper, rhodium, silver, but there's no iron. Iron has never been reported to catalyze this reaction. And look at the activities: at most one hundred total turnover numbers. Nice selectivities, but the activities are pathetic, and a lot of these have to work at minus 80 degrees centigrade. We wondered whether we could find a protein with activity for this reaction.

By now, we have in my refrigerator at Caltech a good collection of hemeproteins that we can test. Interestingly, it turns out that it is not hard to find activity for forming carbon–silicon bonds, as many hemeproteins will do it at some level. The one I like most, however, comes from a thermohalophile, a bacterium from hot springs in Iceland. This cytochrome *c* is so stable that you can autoclave it! It is like a little rock. It's a small protein too, only 124 amino acids. This protein has no native catalytic activity; its function is electron transfer. In fact, its crystal structure shows a histidine coming up to ligate the iron, and it's got a methionine on the other side as the sixth, axial ligand. No free coordination site! If you think too hard about these things you might not even test it. In fact if you try to calculate the volume of the supposed active site, you find it's zero: there's no active site, at least according to the static crystal structure. If you proposed to test this for catalysis, your professor might say, Ah that's dumb, it obviously won't work. But nature does not care about your calculations. We did test it, and it catalyzes the reaction 40 times. And look at that: 97 percent ee, almost perfect enantioselectivity! That's a remarkable result, just for a native protein; all you have to do is provide these synthetic reagents. And, when you do the directed evolution, targeting this axial methionine ligand and other residues that sit where you think the active site is, you can accumulate beneficial mutations and make catalysts that do this reaction thousands of times, much better than any human chemist has been able to do. A single version of this enzyme can make dozens of new organosilicon compounds. We reported in our paper in 2016 twenty new organosilicon products, 19 of which had never been described before (which of course made dealing with the referees a bit problematic, and we had a supplemental information section describing these things that felt about a thousand pages long with all the NMRs etc.).

I'll just point out a few cool things about this enzyme. There's a double bond on this substrate, and I just told you that if you add carbene precursors, the heme proteins can catalyze cyclopropanation. Or you can insert into the N–H bond of this other substrate. And, in fact, the native cytochrome *c* will generate a mixture of products, with no chemoselectivity. But when you evolve it for silicon–carbon bond formation, you see it strongly prefers insertion into the Si–H bond over cyclopropanation or N–H insertion. It 'learns' how to bind the substrate in its active site to make exactly what you screened for. That's what is so amazing about the world of enzyme-catalyzed chemistry: we can get away from processes that make multiple products, one of which you want and the others you don't, because we evolve chemo-, regio- and stereoselective enzymes.

When you look at how this problem was solved, you realize that we are a long way away from being able to do this by design. With a very high-resolution crystal structure we find that those three mutations have completely remodeled the loop that sits over the active iron atom. Residues on the loop have moved out into the solvent, which carves out a whole new substrate binding pocket. Now, if you roll a little ball over the active site on your computer, you can see that these mutations have generated an active site into which the substrates can fit and where the reaction can happen, better than before.

And it works inside of living cells! The catalyst is fully genetically encoded, nothing fancy about it, no new metals. You just take the DNA encoding these three mutations, put it inside bacteria, and they will do the chemistry happily, for sugar. It may annoy some chemists, who are working really hard to do this in the laboratory, that a bacterium will do it for sugar, but I hope that's the future of chemical synthesis.

When I first told this story, before the paper was published, I couldn't talk to the *Science* reporter who was sitting in the front row during my talk. Afterwards, he wanted me to tell him more. I said, No, you know I can't tell you because I want to publish it in *Science*. The reporter still wrote the story and made up whole parts of it! I was delighted when it went all over the world, because his version showed Dr. Spock and Captain Kirk looking for the Horta, life inside of rocks. But that was backwards. I'm not looking for life inside of rocks — I want to put rocks inside of life. I want to put silicon into life and start exploring what it can do. What happens when you make silicon-containing lipids or silicon proteins? How do they behave, and where can you go if you can build life from whole new backbones?

The ideas I have presented today are really very simple. This is the great thing about chemistry and manipulating the biological world. We now have the tools and methods that allow us to go out and explore well beyond where nature would ever care to go, and perhaps we can make life come up with new solutions to deep human problems. By thinking a little bit about how nature solves a problem, using evolution, we can now expand nature's chemistry space into new reactive intermediates, new catalysts, and lots and lots of interesting new chemistry.

In my lab, smart young chemists are looking at life in a new way, thinking, Wow, could I do this new chemistry? I think we'll see an explosion of new chemistry that can be catalyzed by the biological world, with just a little help. This powerful design process of evolution doesn't just optimize, it also innovates. New species come up, new catalysts come up, whole new solutions to problems come up from this process of mutation and selection. We can now marry this powerful evolutionary engineering process with human creativity and 200 years of wonderful chemistry, and solve important problems. It's a fun place to be.

I'll thank the young people who did all of this work at Caltech. I get to go around and talk about it, but they do it. It is truly exciting to see how creativity pops up when people combine different experiences, and I hope that I have given you an inkling of where the future can go.

A Paradigm Shift in Brain Research (1955–1970): Opportunities and Challenges

Seventy Years of Brain Research — Looking Back and Ahead*

Arvid Carlsson[1]

When science began considering the brain from a chemistry rather than electrical standpoint, this marked the beginning of new perspectives on how this complex organ operated. Seminal work relating to dopamine helped to introduce the breakthrough concept of neurotransmitters leading to the understanding of the numerous circuitries within the brain. The observance of decreased dopamine levels when anti-psychotic drugs were administered helped to launch new medical treatments especially for patients suffering from Parkinson's disease. Today stabilizer compounds are being tested to provide beneficial therapies for both elevated or decreased levels of dopamine as well as for other medical conditions.

Keywords: Paradigm Shift; Brain; Dopamine; Neurotransmitters; Stabilizers.

---

*This chapter was based on a lecture delivered by Professor Carlsson at the Molecular Frontiers Symposium in the Royal Swedish Academy of Sciences in May 2014.
[1] Nobel Laureate in Physiology or Medicine (2000).

About Arvid Carlsson

Arvid Carlsson, (born January 25, 1923, Uppsala, Sweden), Swedish pharmacologist who, along with Paul Greengard and Eric Kandel, was awarded the 2000 Nobel Prize for Physiology or Medicine for his research establishing dopamine as an important neurotransmitter in the brain.

Carlsson received a medical degree from the University of Lund in 1951 and subsequently held teaching positions there until 1959, when he became professor of pharmacology at the University of Gothenburg. When Carlsson began his pioneering studies in the 1950s, scientists thought that dopamine worked only indirectly, by causing brain cells to make another neurotransmitter, noradrenaline. Using a sensitive test that he had devised, Carlsson detected particularly high levels of the compound in areas of the brain that controlled walking and other voluntary movements. In animal experiments he showed that depletion of dopamine impairs the ability to move. When Carlsson treated dopamine-depleted animals with the amino acid l-dopa, the symptoms disappeared, and the animals moved normally again. This led to the use of l-dopa as a treatment for Parkinson disease, and it eventually became the single most important medication for the disease. Carlsson's work also contributed to an understanding of the relationship between neurotransmitters and mental states and led to the introduction of new antidepressant drugs.

Although the title of my talk for today indicates that I will be discussing my 70 years of research on the brain, I have reduced this claim by ten years. This because my first ten years of scientific work were dedicated towards training as a pharmacologist at the University of Lund. It was a good training which pretty much focused on the biochemical aspects of pharmacology. However, it was not until a decade later in 1955 when I went on sabbatical that things really started to happen in my research. A paradigm shift in our understanding of the brain was underway and perhaps I should first explain what I mean by this, before expounding on my work.

It was at the time when people first really started to think of the brain in terms of chemistry. Before that, the brain had been considered as an electrical device. And as such, of course truly a black box. And when the brain began to be considered from a chemistry angle it opened up entirely new perspectives. And it was towards the realm that I gravitated. In my opinion this was when the whole early understanding of the brain began. Shortly prior to my entering this field, I was involved in a discovery that would eventually lead to unravelling the role of dopamine as an independent neurotransmitter and along with this, new ways of considering the brain. At the time however I was unaware that this was a breakthrough discovery. And what did this involve? Well, we gave animals in the lab an anti-psychotic drug called Reserpine, not much used anymore, and following this analyzed their brains for serotonin levels, only to find that after a few hours the serotonin had disappeared. At the time serotonin had been identified as present in the brain and speculation relating to its role was that it was important to preserve sanity.

What happened next is that some very clever people, especially in the field of organic chemistry, attempted to explain why the elimination of the serotonin had occurred. Among these was Professor Bernard Brodie, a very remarkable individual who ran one of the top labs in chemical pharmacology at the National Institutes of Health in Bethesda. Work was done to try to explain the pharmacological action of Reserpine in terms of the disappearance of serotonin when the drug was administered, and to make it short, they were unable to explain this. I was there in the Bethesda lab during this time, and as a result was introduced to what were for me entirely new methodologies as we tried to work out the explanation to the mode of action for Reserpine. After this I was slated to return back to Lund, my home university, however even prior to my arrival back I had worked out an agreement with a clever histologist, Nils-Ake Hillarp that we would join forces and continue to work on solving the mechanisms relating to Reserpine. We decided to take a look not only at serotonin but at some other closely-related chemicals — catecholamine, adrenaline and noradrenaline. So upon my return, we began immediately working on this. As a result of this research we discovered that when Reserpine was administered not only serotonin was impacted.

At the top you can see some rabbits. They have been given the anti-psychotic drugs, and it is obvious that there is not much in terms of wakefulness or motility in these animals. We can place

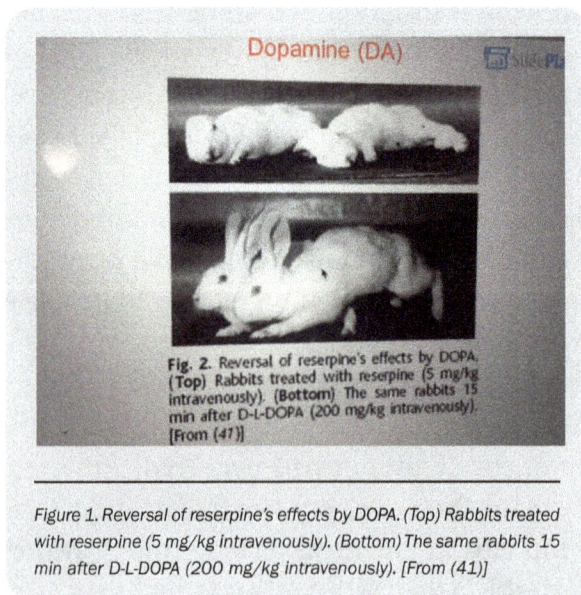

Dopamine (DA)

Fig. 2. Reversal of reserpine's effects by DOPA. (Top) Rabbits treated with reserpine (5 mg/kg intravenously). (Bottom) The same rabbits 15 min after D-L-DOPA (200 mg/kg intravenously). [From (47)]

Figure 1. Reversal of reserpine's effects by DOPA. (Top) Rabbits treated with reserpine (5 mg/kg intravenously). (Bottom) The same rabbits 15 min after D-L-DOPA (200 mg/kg intravenously). [From (41)]

them in these unusual positions and they remain so. However after giving them an IV injection of the precursor of noradrenaline, L-DOPA, they were within a short period up and running again. All of the symptoms brought on by the administration of the Reserpine had seemingly reversed and we realized that it was not the serotonin that had caused this. We thought that it was the noradrenaline. But when we analyzed these brains, we observed that there was still noradrenaline there. So we thought that it had to be dopamine and so we developed methods for analyzing dopamine levels and sure enough, it proved to be dopamine. Dopamine disappeared after this anti-psychotic drug was administered and returned when L-DOPA is dispensed. So that's how we discovered that dopamine occurs normally in the brain as it was not known before.

So that was how the whole notion of Dopamine and the unraveling as to its role began. We were, of course very excited about this discovery. And I remember especially one meeting in London in 1960 with Sir Henry Dale, a Nobel laureate and the top man in the field of neuro transmitters at that time, along with a number of other illustrious scientists. We presented our story and Dale and the others all agreed that our results were in error. They unanimously agreed. And so Hillarp who had also attended the meeting, and I had to return home. We were adamant that we had to do something to convince the scientific community that we were correct in our findings. So Hillarp, an especially skillful histochemist, developed a method by which it was made possible to visualize these compounds — dopamine, noradreline and serotonin in the brain at the cellular level, at the microscopic level. To do this fluorescence microscopy was used and sure enough, we found after florescent treatment, that there were dopamine cells in the brain stem of a rabbit. And with this information we could then map the neural pathways for noradrenaline, dopamine and serotonin. At the synaptic level we were able to come up with a model where we could explain the effect of Reserpine as acting in a storage vesicle. In the transection of the nerve there could be either a dopaminergic, noradrenaline or serotoninergic transmission and we could show how Reserpine acts to block this uptake. There was another uptake from this, with another anti-psychotic drug called chlorpromazine, it did not work at all at this location. However, it worked on the other side of the synaptic cap at the receptor. This work proving our earlier claims was done in 1965, five years after the London meeting. Another international meeting was set up this time in Stockholm to consider the findings and everybody agreed indeed these compounds were important. And we were right, these were the neurotransmitters in the brain.

Then something happened that in parallel with the basic work that we were conducting. When a Viennese neurologist named Walther Birkmayer, who was working in a geriatric hospital in Vienna, heard about the effects of the antipsychotic drugs, in particular that Reserpine caused the depletion of serotonin, he knew from his numerous neurology patients that treatment with Reserpine exhibited a side effect representing symptoms that very much mimicked Parkinson's Disease. And so he came up with a great idea which was to analyze the levels of serotonin in the brain of diseased Parkinson people and see whether or not it had been depleted. Birkmeyer went to the pharmacological society of Vienna and tried to persuade them to do this, but it was deemed a bad idea. Therefore nothing came from this initiative. After we had published our dopamine data however, the pharmacological society changed their minds. An analysis of dopamine levels was conducted and sure enough, the findings showed that the levels of dopamine had depleted. However, if they had conducted the serotonin analysis earlier when Birkmeyer had suggested this, they would have found that serotonin too, although not completely gone, was very much reduced.

So an opportunity was missed, but none-the-less it was this fellow Birkmayer who first came up with the idea of a link between these mechanisms and Parkinson's Disease, the relation to movement control and all that. As a result, he collaborated on a lot of important work in this field and also gave, for the first time, L-Dopa to Parkinson' patients and it worked. He had movies proving that it was effective, however there were a number of other clinics around the world who tried this but the outcome was variable. We really don't fully understand why the treatment didn't work everywhere. Before I go on with a story, I would like to mention how the world responded to all these findings. Take a look at the number of citations, from 1950 through 2000 and beyond. The green curve here is the number of annual citations for dopamine. What you can see is that during the late 50's the years when we did most of the crucial work that was rejected in 1960 in London as well as in the following years, during the period when we conducted additional work to further substantiate our findings, there were hardly any citations at all. This even though there was work being done in Vienna as well as some similar research in Japan, all of which substantiated that we had discovered something terribly important. It took a very long time for widespread acceptance of dopamine as anything significant however, in part as the pharmacologists knew that dopamine as an intermediate in peripheral organs, measured as inactive and therefore it was supposed, not of consequence. This perception was maintained until the late 1960's.

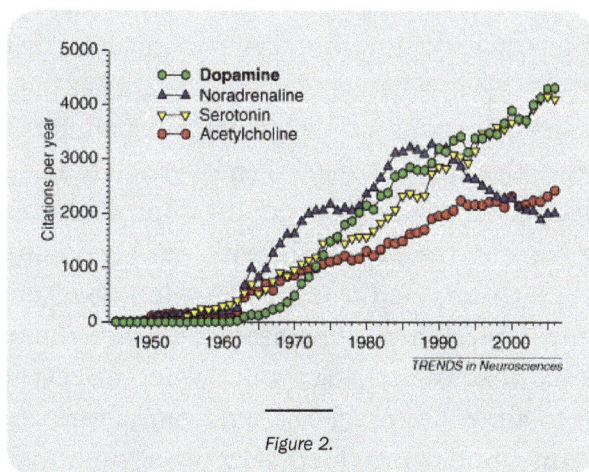

Figure 2.

In 1967, a Greek-American neurologist, by the name of George Cotzias had an idea to change the process by which dopamine was administered. Until then this had been administered by intravenous injection, however Cotzias introduced an oral procedure, which slowly increased the quantity of the dosage and this method proved to work tremendously well. The new therapeutic approach was so effective that within a few years, the treatment of Parkinson's Disease with L-DOPA had successfully spread all over the whole world, which was of course a dramatic event. It was then that the citations started to increase. This fantastic medical story also caught the imagination of the general public through the book written by Oliver Sacks that became the movie titled "Awakenings". In the film the audience is treated to "awakenings" in patients similar to those that we had demonstrated in our rabbits, an exciting experience to witness.

Once the concept of a neurotransmitter had finally been accepted, a number of other neurotransmitters were identified allowing for the making of all kinds of circuitries. I am not going into details here but of course it is very important to really get a better understanding of how the brain works. Here you have the cortex and the striatum where the Dopamine is very much located and you also have the GABA receptors and further down is the brain stem with the dopaminergic and other monoaminergic cell bodies. When trying to understand these complex systems a number of pharmacological mistakes were made early on. These are understandable as what had been discovered at that time was that anti-psychotic drugs act by blocking Dopamine. There was some evidence that there actually is an elevation of the dopamine function within the cervix, however the dopamine present there represents more of the psychosis. This has been confirmed in various ways later on. On the other hand, in Parkinson's Disease you have the depletion of dopamine and other chemicals and then by means of administering agonists, you can reverse this. So initially for treatment of psychosis, antagonists were used. These dopamine antagonists provided very powerful anti-psychotic action. Before this time when we began to first understand the mechanisms for how these were working, the effects of treatments such as Chlorpromazine had been discovered mostly by serendipity.

So you had these findings and the pharmaceutical industry started to work on these. Both agonists and antagonists were very powerful, the idea was to screen for the most potent, the most selective agonists or antagonists that could be found. For those low in dopamine the thinking was that they should receive agonists. Whereas for those with higher dopamine levels, they should be treated with antagonists. This concept was utilized when we began dealing with other neurotransmitters as well. New powerful in vitro screening technologies were being put to use to help with the development of anti-psychotic drugs. With these you could target a specific molecule, a receptor, such as a dopamine receptor or D2 receptor, that you believed could be important for a certain disorder. Using molecular biology one could get these transected into cells or have them multiply. Very powerful screenings were conducted looking for the most powerful antagonists and agonists using this wonderful technique and what resulted was that the strongest of these were indeed found. These developments were useful and led to a number of new compounds. However, the mistake that was made was that the primary screening involved looking for the most powerful option and of course this was the outcome. Certainly progress was made, however if you compare the degree of pharmacological progress compared to the scientific knowledge that was accumulating in this field, the bedside utility was a rather modest one must say. By and large the premise to look right away for the most powerful remedy was a great mistake and I think one of the reasons why the pharmaceutical industry in developing new drugs, insofar as the brain is concerned, hasn't been doing so well lately. Of course, there are other reasons as well for that but the perspective involving the search for the most potent application has been an influence. So, what should be done in order to improve this?

Another thing before I address this question is that what we have learned over the years relating to neurotransmitters is that if you manipulate neurotransmitters you can generate acute, very powerful effects. Either therapeutic effects or side effects, but in addition to this, there are long-term effects. Very strong effects that can really alter the brain and I am now coming in to the concept of plasticity which, of course, has been known for a long time. There is however a growing awareness that this plasticity aspect is enormously important. I will give you one example. Such an experiment or study cannot be done in humans for ethical reasons, so these are being conducted in macaws and monkeys. What we see here is an experiment where you have S that stands for sham control, you have H that's for haloperidol, a powerful anti-psychotic, such as Olanzapine, a more modern type of anti-psychotic, but both of these represent anti-dopaminergic compounds.

They drugs were administered to monkeys in a very careful, controlled way and a series of studies were conducted over about a year and a half and following this their brains were analyzed. One can see that there were even microscopic changes. Evidence that the brain is being rebuilt by these very powerful compounds. So what we see here in this one example, is that there is a reduction of gray matter in certain cortical areas. This is the result of treatment from the drugs administered in comparable protocol to those used in humans. And what we know is that changes such as those that you see here can also be seen in schizophrenia patients and the question is, are these the results of a disorder or is it the drug that causes the change? And it's very hard to tell the difference as this kind of experiment cannot be conducted in humans, of course. A human cannot be given a substance for a year and a half to see if their brain shrinks or not.

So this is the information that we have. This series of studies conducted in Pittsburg, indicated that when you block these particular receptors a profound change occurs. Agonists overall have been abused as drugs. They do terrible things to human beings and, therefore, should be avoided. Although we cannot stop using the anti-psychotic drugs that are available this time, we must try very hard to get rid of them. What we have come up with and

I will try to make this short, is the concept of stabilizers. So this is a stabilizer, these are very important. There are the pros to stabilizers, of course, as you can stabilize something like dopamine for example if it is high, you can bring it down and if it is low you can bring it up. However, there are also cons as of course you are dampening the system and that can change these levels. This compound that is being used to test the stabilizer hypothesis has a preferential action mechanism for receptors that lie along the dopaminergic nerve. Semiological techniques conducted by electron microscope allow for the actualization of the location of the receptors such as the D2 receptors. Some of the receptors are sitting on the dopaminergic neuron itself, while others are in other neurons. Here are located the auto receptors but there are also hetero receptors. As mentioned the stabilizing compound that we are trying now has a preferential action mechanism. In very low doses it starts to act which means that now in a very mild physiological manner you are increasing the dopamine activity very differently from what centrum stimulants, such as what amphetamine and the like are doing. So this activity represents a very smooth physiologic stimulation, a well-controlled stimulation of the dopaminergic system. These compounds are antagonists, in principle. So if you are able to antagonize auto receptors preferentially, you can stimulate dopaminergic activity. But if you give a higher dosage, it starts to work also on the other dopaminergic symbols that you have as well. These will want to be stimulated so that they can operate in the opposite direction. There is a balance effect here.

In animal testing, animals that were placed in a cage with few distractions, do not exhibit much activity or stimulation after they have explored the cage. At this point there is limited activity and when you give an increased dosage of the stabilizer you see in these animals that the dopamine level starts to rise. On the other hand, if you have animals that are introduced into a big arena with lots of things to explore, they start out with very high levels of dopamine and if now you give the same drug in increased dosages, this starts to go down. So this is the stabilizer function, if you a situation that is in between, neither high nor low levels of stimulation, there is almost no effect from the compound. If levels are high, they go down, if they are low, they go up. That is the stabilizer concept. We have demonstrated this repeatedly in a number of animal studies, and this has also been shown to occur in humans as well. And before going there, I just want to just to say that if you conduct a study with the drug in healthy volunteers, where you have a 0.2 micromolar concentration in the blood you have about 20–25 per cent displacement of radioactive erythrocytes you are able to see that there has been a binding to the D2 receptors. In addition to the binding by the drug itself, you also have the displacement in use by the released Dopamine that has been stimulated by the blocking of the auto receptors. So we have a very good understanding as to how these drugs act. It has also proven effective for patients experiencing mental fatigue. For this study people were selected following a stroke. They are in a poor condition, exhibiting an extreme kind of fatigue. The original study and we have conducted several since then, was published in 2012. We found a very clear cut effect. Those who are on the high columns here, are the ones who said that they felt much better on drugs. There was only one fellow who felt the opposite and we think that he was a responder as well. So, there is a very high percentage of respondents amongst these mental fatigue patients. All the studies that have been done all show statistically significant effects. The studies have not been very big but still large enough to demonstrate statistical significance. This also includes one study where you do have not a deficiency of Dopamine, but rather an agonist effect on the Dopamine due to alcohol intake. The study conducted at the Karolinska Institute also considered effects on alcohol cravings upon treatment with this drug and reduced cravings were confirmed as statistically significant. I think we have reached a proof of concept here and that these drugs and this principle of using stabilizers rather than full blown agonists and antagonists has a great future, and I thank you so much for listening.

CRISPR System:
From Adaptive Immunity
to Genome Editing*

Jennifer Doudna[1]

CRISPR-Cas9 is a revolutionary genome-editing tool. Understanding how Cas9 recognizes DNA and how to control its function will be critical in improving the system. We used single-molecule FRET to elucidate a key validation step during DNA target recognition. We also used X-ray crystallography to show how a Cas9 inhibitor is able to permit DNA binding but prevent cleavage. Finally, CRISPR research is notable not just for the exciting applications, but also for its profound ethical implications.

Keywords: CRISPR; Genome Editing; Biochemistry; Structural Biology; Bioethics.

---

* This chapter was based on a lecture delivered by Professor Doudna at the Molecular Frontiers Symposium in the Royal Swedish Academy of Sciences in May 2017.
[1] Howard Hughes Medical Institute, University of California, Berkeley, CA 94720-3220, USA.

About Jennifer Doudna

Jennifer Anne Doudna (born 19 February 1964) is an American biochemist, professor of chemistry at the Department of Chemistry and Chemical Engineering, and Professor of Biochemistry and Molecular Biology at the Department of Molecular and Cell Biology at the University of California, Berkeley. She has been an investigator with the Howard Hughes Medical Institute (HHMI) since 1997. She directs the Innovative Genomics Institute, a joint UC Berkeley-UC San Francisco center and holds Li Ka Shing Chancellor's Professorship in Biomedicine and Health, and is the chair of the Chancellor's Advisor Committee on Biology at UC Berkeley.

Doudna has been a leading figure in what is referred to as the "CRISPR revolution" for her fundamental work and leadership in developing CRISPR-mediated genome editing. In their seminal 2012 paper "A programmable dual-RNA-guided DNA endonuclease in adaptive bacterial immunity", Doudna and Emmanuelle Charpentier were the first to propose that CRISPR/Cas9 could be used for programmable gene editing, which is considered as one of the most significant discoveries in the history of biology. Their work has since been further developed by many research groups for applications ranging from fundamental protein research to treatments for diseases including sickle cell anemia, cystic fibrosis, Huntington's disease, and HIV.

Doudna has made fundamental contributions in biochemistry and genetics and received many prestigious awards and fellowships including the 2000 Alan T. Waterman Award for her research on ribozyme, and 2015 Breakthrough Prize in Life Sciences for CRISPR/Cas9 genome editing technology (with Charpentier). She has been a co-recipient of the Gruber Prize in Genetics (2015), the Canada Gairdner International Award (2016) and the Japan Prize (2017). Outside the scientific community, she has been named one of the "Time 100: Most Influential People of 2015" (with Charpentier) and listed as a runner-up for "Time Person of the Year" in 2016 alongside other CRISPR researchers.

So, what we're going to hear about in the next two talks is research that I think has a lot of themes that we heard about this morning, a fundamental discovery, curiosity-driven research, scientists who were working collaboratively internationally together on a project that was aimed at answering a particular question, as I'll tell you, but ended up in a very different direction from where we expected it to go in the beginning. And, you know, I think those are the themes that I really have always valued myself as a scientist. And especially, for the students that are in the audience, I wanted to tell you a little bit about my own background, how I came to do what I'm doing and how fortunate to have the opportunity to do science.

So, you know, I think the story really begins with thinking about how information is encoded biologically. You know that we're all kind of comfortable now with computers and how information is encoded electronically, but in cells, of course, information is encoded chemically and its encoded in DNA. Scientists have understood for decades that this molecule, the double helix is really about giving the instructions to the cells or organisms that allow them to divide, to develop into certain types of tissues or whole organisms. And that by understanding what that information is and how to manipulate it, we can have a lot of ability to control biological systems, and we already heard a lot about that in Francis Arnold's talk.

And for me, when I was growing up, I grew up in a small rural town in Hawaii. Nobody in my family was a scientist, but my dad was a professor of American literature and he loved shopping in used bookstores. And so, one day I came home from school, I was probably in sixth grade and he had left a copy of *The Double Helix* on my bed, sort of a bent, dog-eared paperback copy of Jim Watson's famous book. And it..I just...I picked this book up and I had no idea what it was about. And I read it and it kind of blew my mind because it was really about the story of how scientists had figured out the structure of the double helix, how they had done experiments to actually understand the structure of a molecule as important as this. And I think for me when I think back on it, that was kind

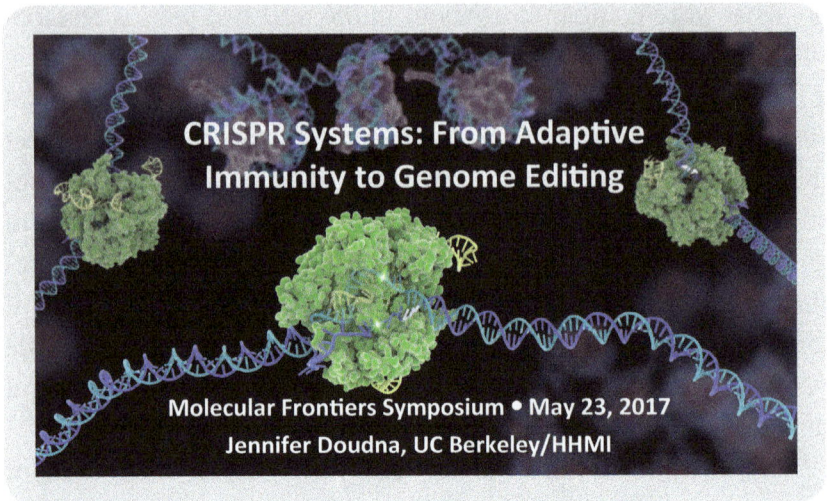

of the moment when I really thought to myself, wow I wonder if I could do something like that someday, not something that momentous of course but just how could I have the opportunity to do science, could I do research, could I, you know, solve puzzles in biology or chemistry by asking questions and then answering them experimentally. It seemed like a really exciting possibility. So, as I went along in my career, I went to college, I majored in chemistry with a focus on biochemistry. And then I had the good fortune to go to graduate school and join the laboratory of Jack Szostak. When I got to Jack's lab, there was a lot of work that had gone on in his lab in the past about genetic recombination and the way that the ends of chromosomes called telomeres are structured and how they're replicated. But, that was kind of in the past for him and he was really moving in a different direction. He was thinking about the origin of life, a lot of the questions that he talked about this morning. And so, when I joined his lab, I had chemistry background, I didn't know too much about biology really, but we thought a lot about sort of this what we call the central dogma, how information is encoded and how it's passed along from generation to generation biologically and how that feeds evolution and then if we look back

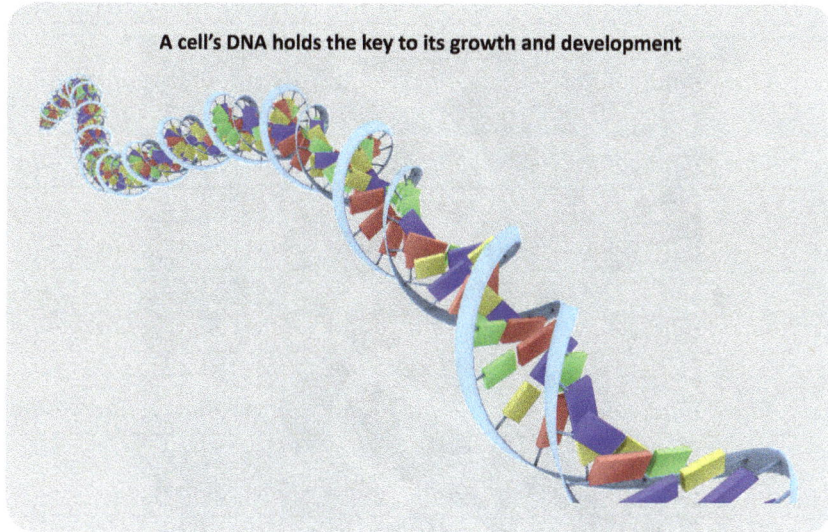

A cell's DNA holds the key to its growth and development

in time, can we look at what's happening in modern biology and understand where it came from. And so, this is really just showing a very simplified view of the central dogma, but fundamentally information is encoded in DNA. It gets transcribed into molecules of RNA, which are sort of I call it DNA's chemical cousin, as it has the same sort of polymeric structure of four individual letters that make up RNA molecules. And then, if these RNA molecules are messenger RNAs, they actually encode proteins that the cell will then recognize as a coded message that tells the cell how to make a certain type of protein. And so, all of these proteins then get made and they carry out most of the functions in the cell. But, when I was starting graduate school, one of the things that was really, really exciting and kind of mind blowing at the time was that scientists, including Altman and Chuck and Norman Pace and a few others had figured out that there are examples in biology of RNA molecules that have function in their own right. They're chemically active, they can function as catalysts. And this was something that had caught Jack's attention, in particular, because of thinking about the origin of life and could these kinds of catalytic RNAs that we call ribozymes

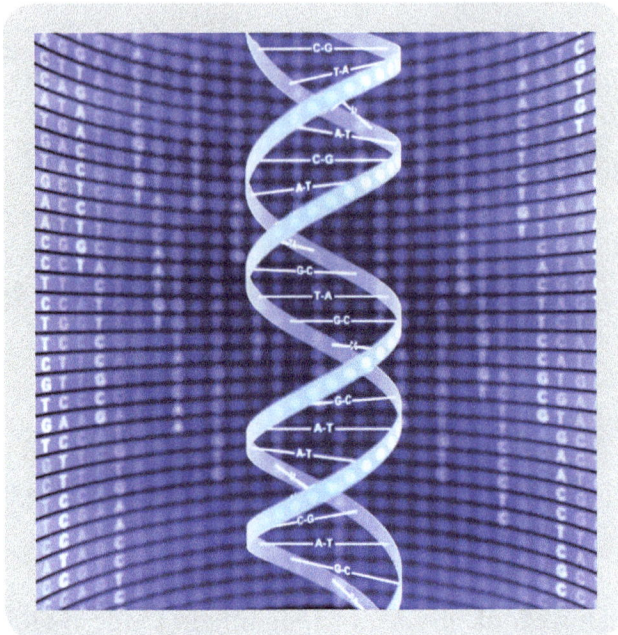

could they actually tell us something about early evolution. So that was a project that I started working on as a student.

And so, really, when I...over the course of my career, I've continued with that sort of central theme of really thinking about the function of RNA molecules in biology and in particular, RNA molecules that don't encode proteins, but actually where the buck stops here. In other words, this molecule has a function in its own right that is important for biological systems. And...early in my career when I went on to do my postdoc in Boulder, Colorado, in the lab of Tom Cech and then later when I started my own lab at Yale University, the first few years of my work were focused on figuring out the molecular structures of these kinds of RNAs that are functional as RNA, you know, functional as catalytic RNAs in particular. And it was really through that line of research that I eventually started thinking about how to link that more chemically and structurally focused line of research to what happens in modern biology. And in particular, all of the work that was going on around

The "central dogma"

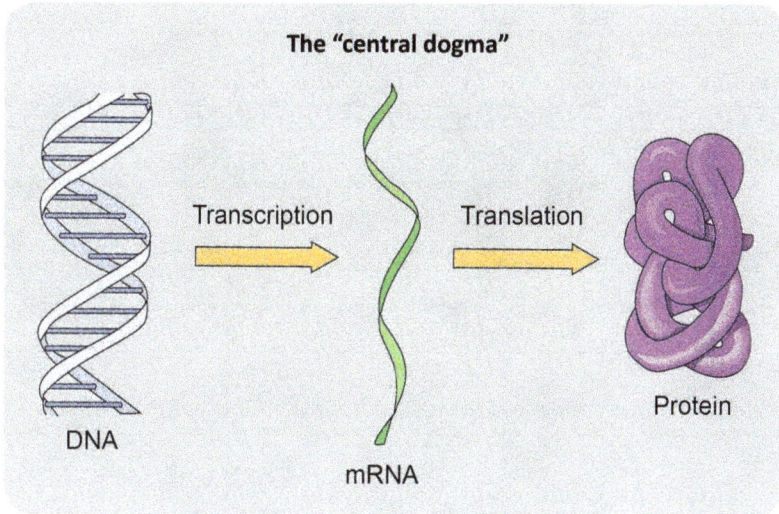

Transcription

Translation

DNA

mRNA

Protein

me at the time where many people were discovering more and more examples of RNA molecules that do very interesting things to control the way that information is expressed in cells. In other words, how DNA, the information encoded in DNA is actually deployed in the cell. Turns out that RNAs…in many…there are many really interesting examples now where RNA molecules are really the mediator of that kind of deployment of information.

And in particular, we started working on small RNAs called microRNAs that are made in many types of animal and plant cells, cells that we call eukaryotic cells that are used to control the flow of information and working on the molecular mechanisms by which those microRNAs are produced in cells and how they are actually able to interact with molecules, other RNA molecules in cells and control the way that proteins are made.

And why am I telling you all of that? Well, that's really the precursor to the story of CRISPR, for me at least, because it was through that early work in my lab, once I moved my lab to UC Berkeley in 2002 that I got acquainted with a colleague there Jill Banfield who was working in a completely different college in a different field. She is not an experimental biologist, she does only

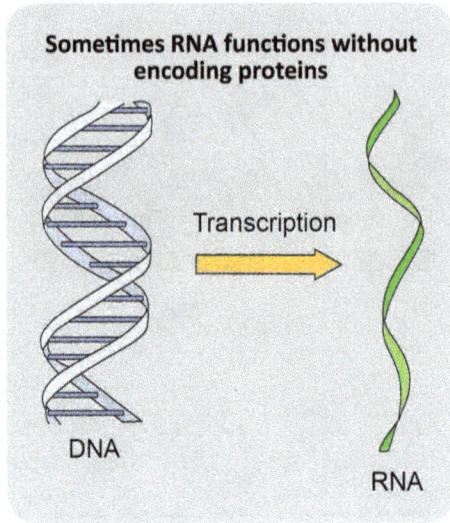

computational work. But, her lab was working away on bacteria that are found in various interesting environments. And she would go on these crazy trips to Australia and the outback of California even and isolate samples of bacteria that were growing in pools that were very acidic, pH1 and or very low or high temperatures and try to figure out what kinds of bugs were growing there. And the way they did that was simply to sequence all of the DNA in those samples and then piece back together the bacterial chromosomes, which are illustrated by this circular drawing here. And then from that sequencing information, figure out what kinds of bugs were growing in those environments. And importantly also, the kinds of bacteriophage, the kinds of viruses that would infect those cells because it turns out that just like us and virtually every other form of life that we know on this planet at least, there are viruses that can infect cells and take over their machinery for making new viruses.

So, Jill Banfield was doing this research and I didn't know anything about it. And then one day, she called me because she said, you know, we've come across something very interesting in our work and it's very puzzling. And I think as an RNA biochemist, you might be someone who can help us. And I thought, really?

CRISPRs: Hallmarks of acquired immunity in bacteria

<u>C</u>lusters of <u>R</u>egularly <u>I</u>nterspaced <u>S</u>hort <u>P</u>alindromic <u>R</u>epeats (CRISPRs)

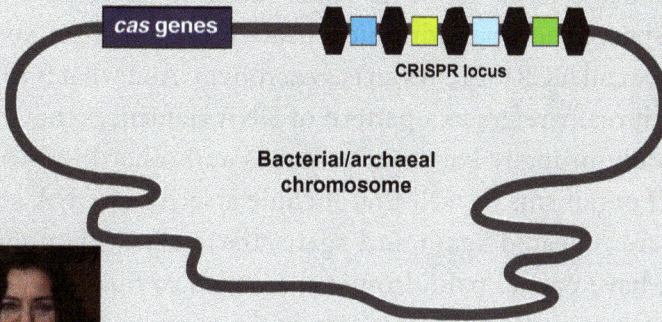

cas genes

CRISPR locus

Bacterial/archaeal chromosome

Jillian Banfield, UC Berkeley

Bolotin *et al.* 2005
Mojica *et al.* 2005
Pourcel *et al.* 2005

CRISPRs: Biology and Technology

- CRISPRs: RNA-guided genome protection

- Anti-CRISPRs: Phage fight back!

- Ethics: Responsible progress

CRISPRs: Biology and Technology

- CRISPRs: RNA-guided genome protection

- Anti-CRISPRs: Phage fight back!

- Ethics: Responsible progress

And, so we got together and what she told me was that her lab and really just a couple of other labs around the world at that time were uncovering more and more examples of sequences that had become known as CRISPRs which stands for clusters of regularly interspaced short palindromic repeats. A big mouthful, you can see why we call it CRISPR. It's an easy acronym. And what is a CRISPR? Well, it really refers to a pattern of DNA sequences that are found naturally in many bacterial genomes and other kinds of single-celled organisms as well that include a series of DNA sequences that are repeated again and again, these black diamonds. They often have partial palindromic characters so they can be read in the forward and reverse direction in the same way. So they have the possibility, if you look at the letters of DNA, they have the possibility of base pairing to form a cruciform sort of structure. And really importantly, in between the repeats in these loci in the genomes that Jill was studying were sequences that were unique and were found in these three papers published in 2005 from three different bioinformatics teams to correspond in many cases to sequences found in viruses. And not only that, they were found in viruses that infect these organisms. And that's what Jill's research had shown a connection to. And her work also shows that these sequences evolve over time. So if you isolate bacteria from a certain site, somewhere in some acid mine tailing or something like that, and then you sequence it and then you go back two weeks later, isolate the sample again and sequence it again, you can find in many cases that now these CRISPR loci have changed and they've changed in a very particular way where new inserted sequences, new examples of these colored boxes have been integrated into the CRISPR locus. So it really looked like this was some kind of evolving system. And the fact that these integrated sequences targeted viruses hinted that this might be some kind of adaptive immune system, a way that bugs could acquire immunity to bacteriophage by storing a little piece of genetic information from each virus that infected the cell

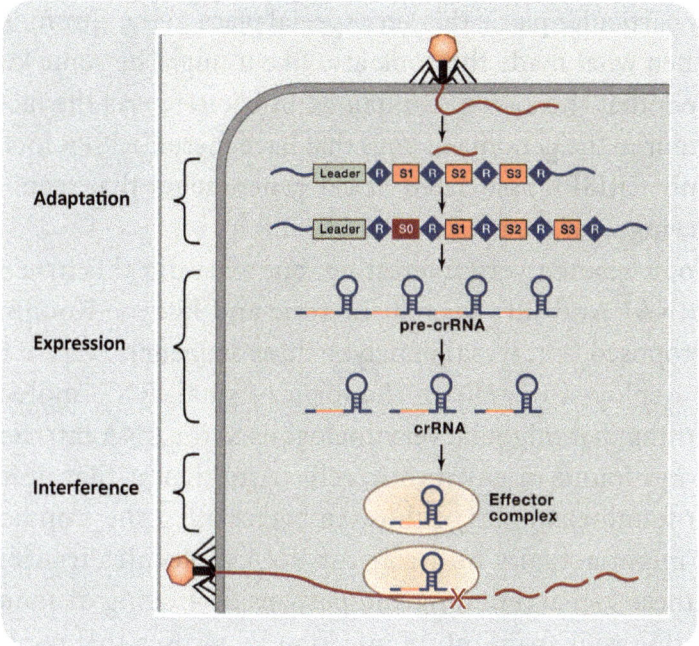

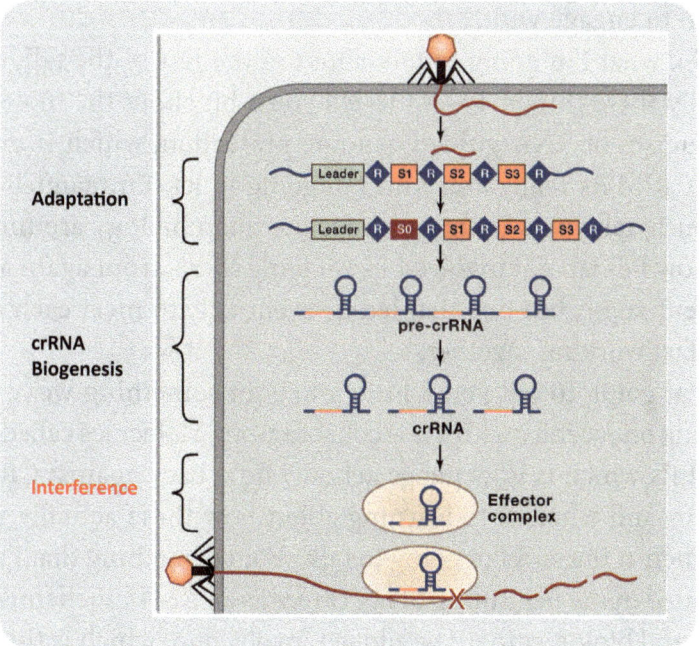

in this particular place, this very special place in the chromosome. And then what made this look also like it might be some kind of a system that was being maintained in bacteria was the fact that next door in the genome of bugs that have these CRISPR loci were typically CRISPR associated or cas genes, genes that seem to be co-varying with the presence of CRISPR loci.

So, it seemed very interesting, but what did this have to do with RNA? And Jill's idea at the time and Eugene Koonin had also proposed this, was that maybe these sequences were actually being deployed in cells in the form of small RNA molecules. Something that might be very analogous to the RNA interference pathways found in eukaryotic cells. And that at that point we hadn't found examples of those in bacteria. So she wondered if there might actually be a role for RNA molecules transcribed from these loci and finding and perhaps destroying or somehow protecting cells from future infection by viruses that contained matching sequences. It was sort of a really fascinating idea. So I was hooked. It would seem like a really interesting problem for my lab to engage with.

So, what I'm going to talk about in the rest of the talk today is really three things. I want to tell you a bit about the molecular mechanism of RNA-guided genome protection, which is exactly what CRISPRs really are. And I'm going to let, Emmanuelle will obviously tell you a lot more about the microbiology around this and how her lab got involved in studying these, from again a very different angle, but one that led us eventually to meet each other and start working together.

I'm going to tell you a little bit about something we've been working on recently with a few collaborators. Molecules called anti-CRISPRs which is how phage actually fight back against CRISPR systems and what we're learning about how these actually work. And then at the end I'm going to talk about something that I never imagined in my beginning of my career as an RNA biochemist and structural biologist that I would get involved in, which is thinking

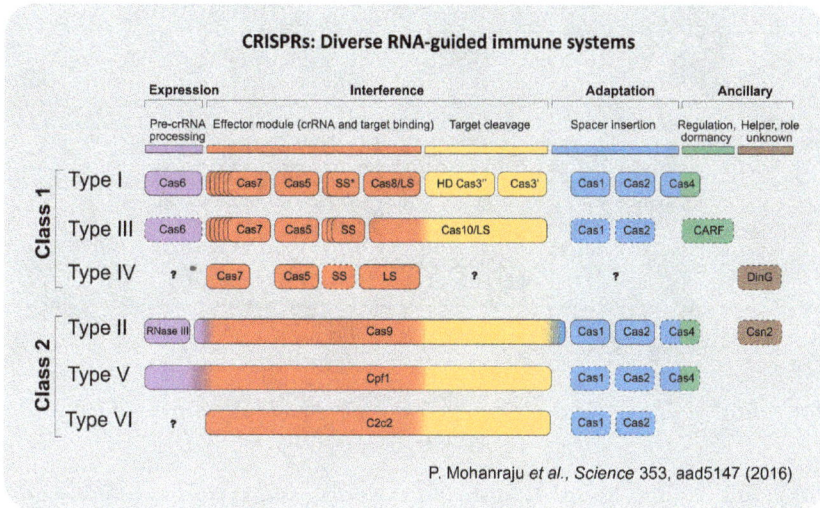

CRISPRs: Diverse RNA-guided immune systems

P. Mohanraju *et al., Science* 353, aad5147 (2016)

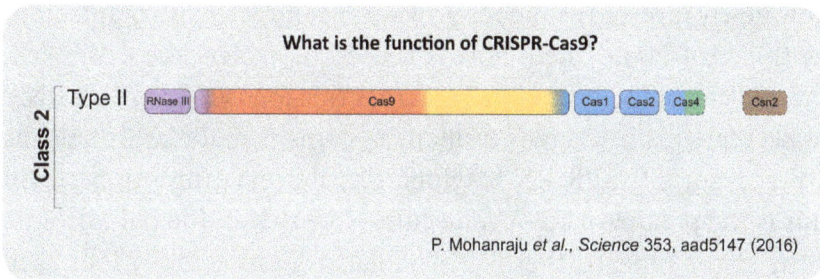

What is the function of CRISPR-Cas9?

P. Mohanraju *et al., Science* 353, aad5147 (2016)

about the ethics of technology that came out of work that had been done in my lab with my collaborators and how we could work together as scientists to encourage what I call responsible progress. So, but first, let's talk about CRISPRs.

So, this is a cartoon that illustrates sort of what we understand currently about the pathway of these adaptive immune systems. And we can kind of break these down into three stages of immunity. So, the first step is that a cell, here's the membrane of a cell, gets infected by a virus. And you can see this phage has landed on the surface of the cell, it's injecting its DNA into the cell. And then if this lucky cell has a CRISPR system that's active, it can actually acquire small bits of this foreign DNA and integrate it into the

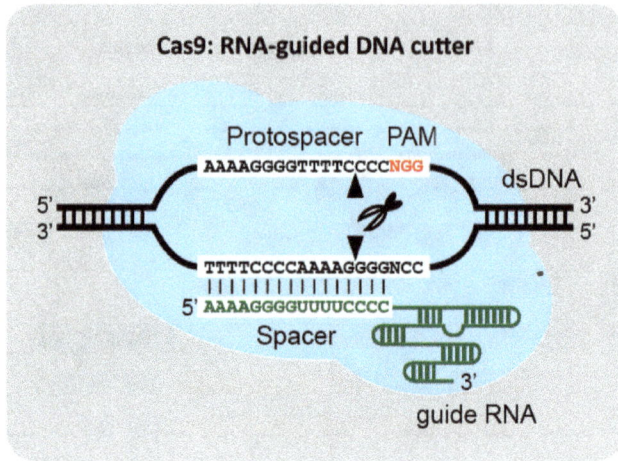

Cas9: RNA-guided DNA cutter

Protospacer   PAM

AAAAGGGGGTTTTCCCCNGG

dsDNA

5'   3'
3'   5'

TTTTCCCCAAAAGGGGNCC

5' AAAAGGGGUUUUCCCC

Spacer

3'

guide RNA

CRISPR locus here in a process that we call adaptation. So this now newly inserted sequence provides a genetic record of infection by this virus. And then, how is it used to protect the cell? Well, the cell can actually transcribe across this locus to make an RNA molecule. This initially is in many systems is made as initially as a precursor CRISPR RNA, which I'm abbreviating pre here. So this is the precursor RNA molecule. Note that in the palindromic parts of the repeats, these RNA molecules can actually fold back to form little hairpins and we know that's actually important for how they're recognized sort of tags the RNA in the cell and tells the cell this is a CRISPR RNA. And then those RNAs are chopped apart and used to generate individual molecules that each includes a sequence derived from a virus. And those RNAs then assemble with one or more of the proteins encoded by the cas genes to form effector complexes. These are the surveillance complexes that can go searching through the cell looking for a sequence of nucleic acid that matches the sequence in the CRISPR RNA. This piece right here. And when that occurs then that effector complex is recruited to the nucleic acid through base pairing and allows the associated cas proteins to cleave the viral DNA. So it's a really, really interesting way that bacteria have evolved to adapt to viruses and fight back.

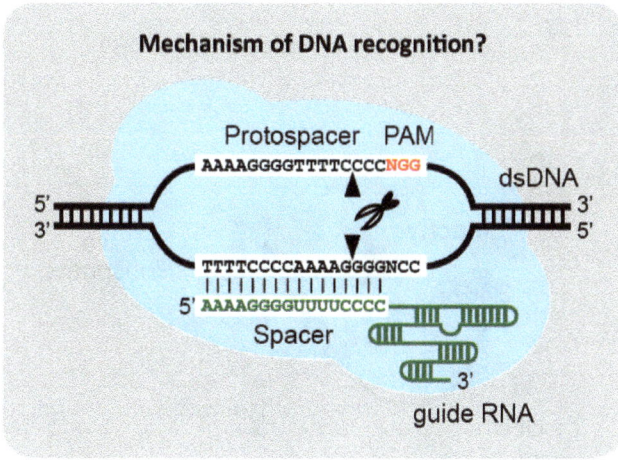

So, in my lab over the last, gosh almost little over 10 years now, we've been working on these systems and we've really been studying all aspects of this, the adaptation step, the expression and processing of RNA and then the way that interference works here, but today, I'm just going to really focus on this step here, which is interference and how these systems are able to protect cells and how an understanding of the mechanism of that led to recognition that this could be harnessed for a very different purpose, namely for genome editing.

So, one of the things that's really cool about CRISPR systems, and I think Emmanuelle might say what about this is that they're really diverse. And this is actually a slide that's taken from a review that was published last year. And what this is showing you in cartoon form is a diagram of the number of genes, number of cas genes that are part of different CRISPR systems that are organized here. And we can very roughly divide these into class I in the top and class II. And the way they differ from each other is that the class I systems all include multiple cas proteins that we know have to come together with the CRISPR RNA to form these targeting complexes. So there we've done a lot of work on these in my lab over the last few years and others have too and we know that these

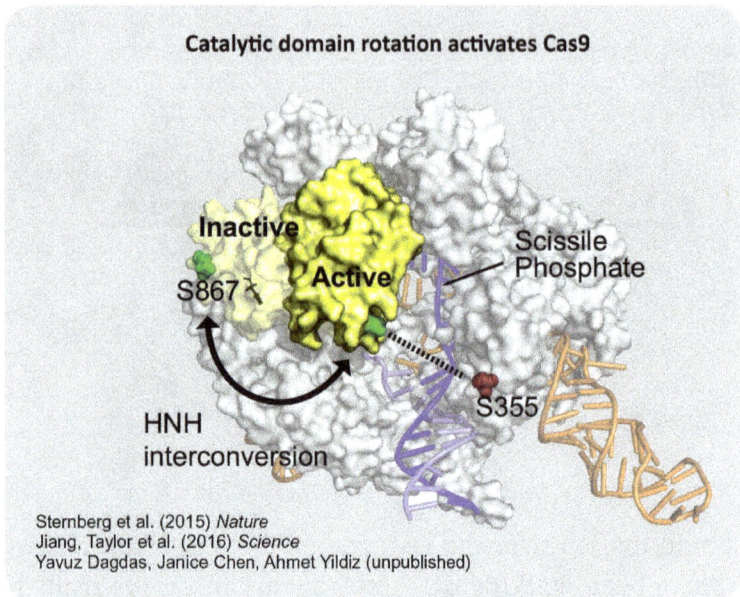

Catalytic domain rotation activates Cas9

Sternberg et al. (2015) *Nature*
Jiang, Taylor et al. (2016) *Science*
Yavuz Dagdas, Janice Chen, Ahmet Yildiz (unpublished)

are now big assemblies, they're very beautiful if you look at the structures and they work by using the information in the CRISPR RNA to find and ultimately to cut foreign nucleic acid. But down here these systems that are called class II are set up differently in the sense that in each case these class II systems have a single cas gene that is responsible for RNA-guided genome protection. And so, initially, we were at my lab, we're sort of focused on, actually working mostly on the type I systems up here in the early days of CRISPR in my lab. And then in 2011 I went to a conference at the American Society of Microbiology and I met Emmanuelle. And she was working on an organism called Streptococcus pyogenes. And in her research on the pathogenesis of this organism, she had been studying all of the small RNAs that are made in this bug and that had led, and I won't spill here sorry, but that had led to some very interesting discoveries about how the CRISPR RNA and another RNA called Tracr are produced in those cells and had led to a very nice publication that she had put out earlier that year.

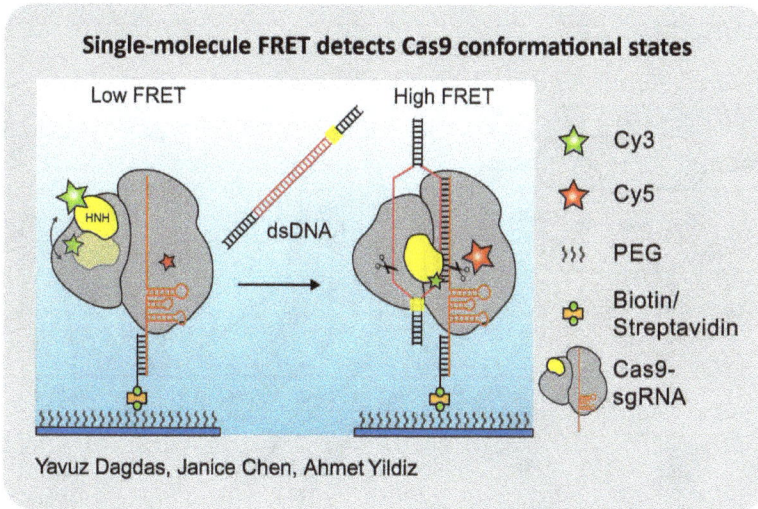

Single-molecule FRET detects Cas9 conformational states

Yavuz Dagdas, Janice Chen, Ahmet Yildiz

So, when we met at the conference we decided to team up to work on a very, what seemed at the time like a really just fun, interesting question, which was what is the function of this protein called Cas9. At the time it was actually called Csn1. So it's had a name changed since then. But this protein was part of the CRISPR system found in Streptococcus pyogenes and it's a great example of systems that are known as type II that have a single gene called Cas9, that is responsible for protecting cells from viral infection in bugs that have this type of CRISPR system. And at the time, there was genetic evidence that this gene was somehow responsible for that protection, but nobody knew the molecular function of the protein that it encoded. And that was the question that we set out to address. And that led to a really, really fun collaboration that was going on across thousands of miles. Her student, Christoph Zielinski was in Vienna, Emmanuelle herself was here in Sweden at Umea University and my lab in Berkeley, California had Martin Jinek, former post-doc who was working on the project from our side. And so Christoph and Martin teamed up to do a series of biochemical experiments that led to the realization that Cas9 is

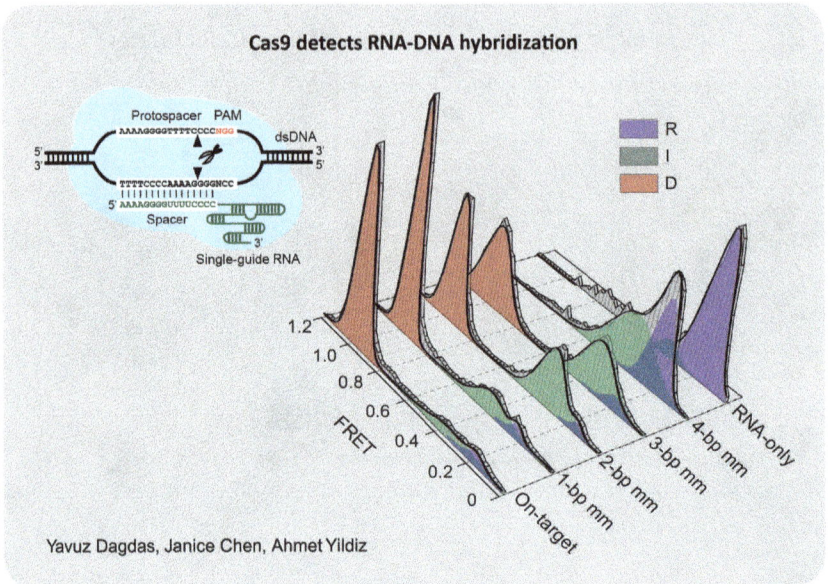

Cas9 detects RNA-DNA hybridization

Yavuz Dagdas, Janice Chen, Ahmet Yildiz

an RNA-guided DNA cutter. And so what I'm showing you in this cartoon is an illustration, a very low resolution illustration of Cas9, this blue molecule, that is the protein that protects cells using this RNA-guided mechanism. And we found out that the way it works is it uses an RNA molecule. In nature it's actually two separate RNAs, a CRISPR RNA (crRNA) that contains the targeting sequence which in bacteria would come from a virus, and then a separate RNA called tracrRNA that base pairs with the end of the CRISPR molecule to form a structure that is recognized by Cas9. And so this dual RNA interaction binds to Cas9 protein to form the functional complex that can interact with DNA molecules. And when that interaction occurs through base pairing between the RNA and one strand of the DNA, the protein is able to unwind the DNA inside the protein and trigger a double stranded DNA break using two separate active sites in the enzyme that cut each strand of the DNA. And Martin Jinek was able to figure out. He made mutations in the active sites of the enzyme and figured out which one cuts which strand of the DNA. And furthermore, that you could actually make mutations

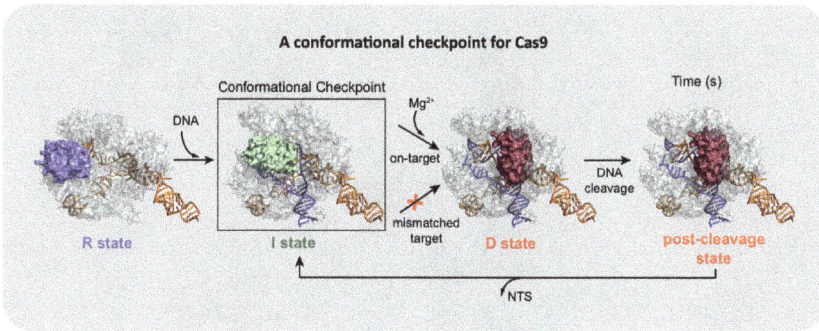

in those active sites and maintain the ability of Cas9 to have this RNA-guided DNA binding ability, but with mutated active sites, it would not cut the DNA and we thought that could be something that seemed like a really interesting feature that the chemistry of DNA cutting was entirely separate from the RNA-guided activity of the protein. We also figured out that there's a little motif in the DNA next to the site that's targeted called the PAM which for this enzyme is a GG dinucleotide. And this was all figured out by using a reconstituted system where we had purified protein, purified RNA. And that led to the realization that we could actually simplify the system compared to what nature has done by linking together the CRISPR RNA and the tracrRNA to create a single guide form of the RNA that would work for this kind of DNA recognition and cutting. And I think for us it was really when we had that, that result and showed that we could get this to work as a two component system with a single Guide RNA (sgRNA), that we realized that there was a very interesting and very different application of this that could be really exciting from a technology perspective.

So, let me show you a little video. It was also made by Janet Awasa, the person who also made the video that Jack showed this morning, that illustrates how these systems work. So let's see.

So here we have some bacteria that are being invaded by a bunch of viruses. And when a virus infects the cell, of course, here it is injecting its DNA, here is the chromosome. And if this all has

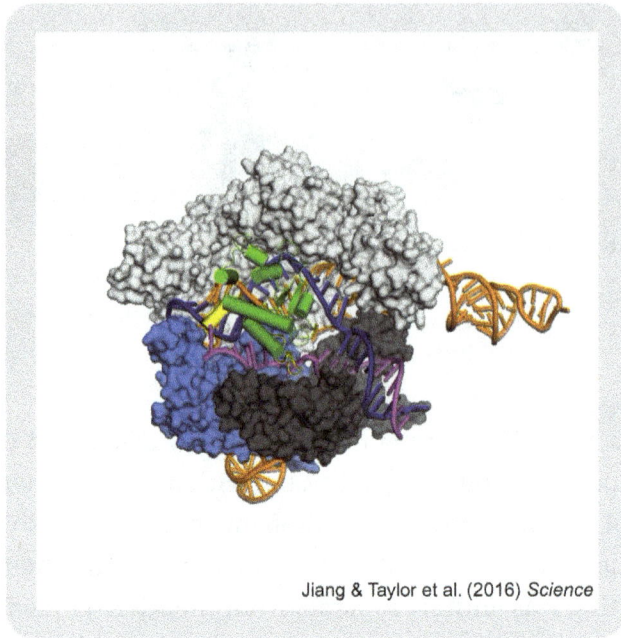

Jiang & Taylor et al. (2016) *Science*

a CRISPR locus, it's able to acquire a new piece of the foreign DNA into the CRISPR sequence by integrating it directly into this array. And as I showed you in the cartoon before, this array is then copied into a molecule of RNA that gets chopped into individual units that each includes one of the virally derived sequences. Those molecules combine with the tracrRNA this separate molecule that interacts at the end of the CRISPR molecule to form a binding site for Cas9. And then together, this RNA protein complex goes out searching around the cell to find examples of nucleic acids that match the sequence in the CRISPR RNA. And if a match is found, then this protein RNA complex can assemble with the DNA, unwind it, and allow an RNA-DNA hybrid to form inside. It actually happens inside the protein. And then there's cutting of the two strands of the DNA that in bacteria leads to degradation of the viral DNA. So, great way to destroy viral DNA, but it turns out from research in many labs over the years including in Jack's, that when animal or plant cells, eukaryotic cells experience a double-stranded break

CRISPRs: Biology and Technology

- CRISPRs: RNA-guided genome protection

- Anti-CRISPRs: Phage fight back!

- Ethics: Responsible progress

Many phages encode Cas9 inhibitors

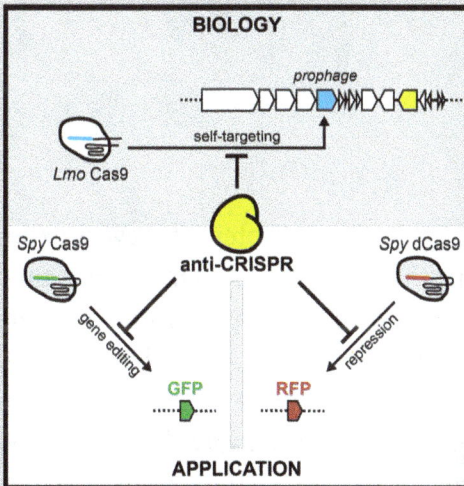

BIOLOGY

prophage

self-targeting

Lmo Cas9

Spy Cas9

anti-CRISPR

Spy dCas9

gene editing

repression

GFP

RFP

APPLICATION

Rauch B *et al.* and Bondy-Denomy, J. (2017) *Cell*

to their DNA, they can repair the break by introducing changes at the site of the break. And so, people had appreciated that if you could introduce double-stranded breaks into genomic DNA at a particular place, that would trigger this kind of repair and you could use this for very precise genome engineering and there was a whole line of research that had led to the development of programmable proteins to do this. And those technologies while exciting had not really been widely adopted because of the difficulties of protein

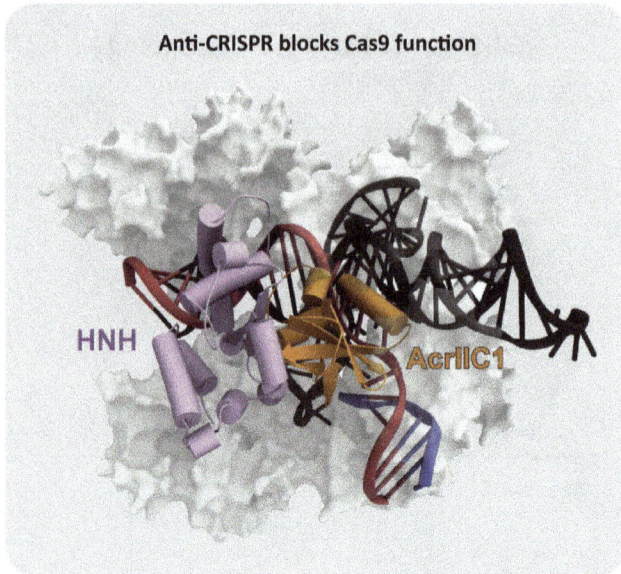

Anti-CRISPR blocks Cas9 function

HNH

AcrIIC1

engineering having to make new proteins for every experiment. We realize with Cas9 that the recognition of DNA works fundamentally differently because it requires not a protein DNA interaction, but an RNA-DNA hybridization event, right here. And that meant that you could use the same enzyme for…essentially for any kind of genome editing experiment by simply changing the sequence of the Guide RNA to match a sequence that was at a desired site where you might want to reduce the change in genome. And so that was the idea that we published when we published this work together with the Emmanuelle's lab in the summer of 2012. And so, you know, it's been really exciting over the last five years to watch this technology taking off and of course we, our labs, have been engaged in various aspects of this. And what I thought I would tell you about for the next few minutes and then a little story that I want to tell you about current science we're doing is about our efforts to really understand at a mechanistic level how this works. And why do I care about this? Well, I think it's, first of all, just a fascinating question to think about this mechanism of DNA recognition because you think about it, this is an amazing enzyme that actually has to melt 20 base pairs or maybe a little bit more than that of DNA. And, so DNA has very stably base-paired the two strands, double helix held together, it's very thermodynamically stable, but somehow this enzyme has to melt it open to allow this RNA-DNA hybrid to form. Not only that, it doesn't have any external energy source to do that. So it doesn't hydrolyze ATP which is a classic way that biological enzymes will recruit or sort of transform chemical energy that they can use for other kinds of work. This enzyme doesn't do that and so we wondered how this kind of melting works, and furthermore what that mechanism could tell us about the way this works in…in actual cells and how we might think about using it for other kinds of technical or technological applications.

So, one of the things that's emerged over the last few years is that there's actually a very interesting structural change that happens in this protein as it assembles with nucleic acid. And this

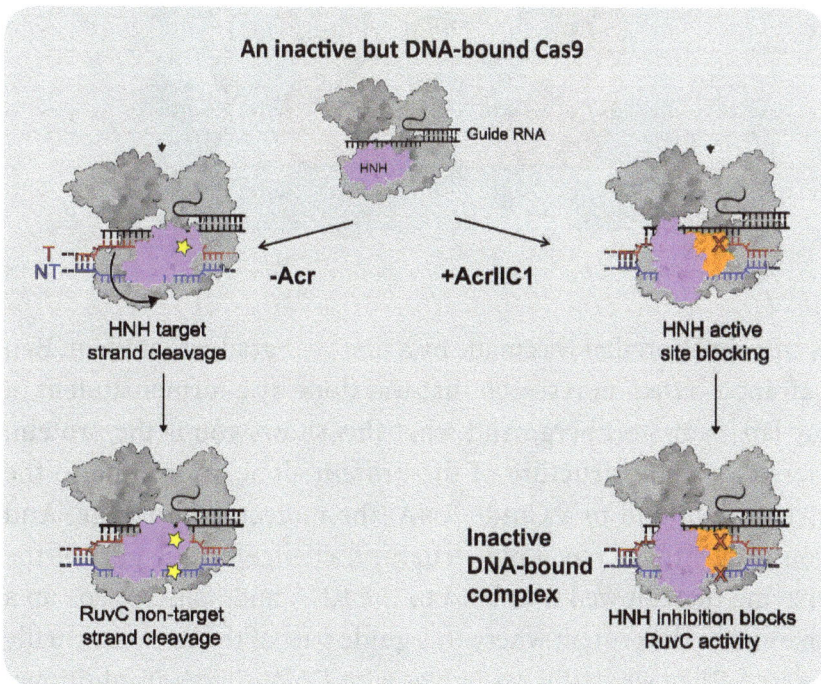

An inactive but DNA-bound Cas9

Guide RNA

HNH

-Acr   +AcrIIC1

HNH target
strand cleavage

HNH active
site blocking

RuvC non-target
strand cleavage

Inactive
DNA-bound
complex

HNH inhibition blocks
RuvC activity

is an animation that was made by a first year graduate student, Ben Lafrance, based on research that was done by a former student in my lab, Sam Sternberg. And what this shows you is the protein. Here's a crystal structure of the protein alone, morphing to the structure bound to a Guide RNA, the molecule in orange. And you notice there was a big structural change in this part of the enzyme that allowed it to bind to the RNA and it also opens up a channel in the protein where this guide part of the RNA is actually located. Then when this assembles with DNA, there's an additional structural change that accommodates the RNA-DNA hybrid that forms. And that seemed very interesting and potentially related to this unwinding mechanism, but we also found that there has to be an additional structural change here, in this part of the enzyme, a domain in the protein called HNH that swings it into place. So it can actually cleave the targeted strand of the DNA. The strand that's actually base-paired with the RNA. And that structural change was not seen in the early crystal structures that were done by various labs because people were using a single strand of DNA to form this hybrid, but without having an actual double-stranded DNA substrate, which is the natural substrate for this enzyme. And so, once it was possible to crystallize, sort of trap that double-stranded form of the DNA on the Cas9 enzyme. Again, here's the orange Guide RNA sticking out, here is the blue and magenta DNA double helix entering the protein and then getting unwound inside the enzyme. We found that in this structure, this HNH catalytic domain is now actually very close to where it needs to be to cut the DNA. And there are lots of reasons for that, that have to do with the way this protein nucleic acid interaction occurs in the center part of this enzyme. So, it's clearly a very interesting little machine that's set up to interact with nucleic acid and really recognize only the Guide RNA that's meant to bind to and then to interact with DNA molecules to trigger only once it's engaged on a DNA that is based-pairing to the Guide RNA. The adoption of this active structure of the enzyme that's now catalytically competent to cut the DNA.

And so, one of the things that we've been studying is this work was initially based on just modeling of what we knew sort of had to happen in the protein. It was very satisfying to see that structural rearrangement once we had crystallized a trapped structure of Cas9 on a double-stranded DNA substrate, but we really thought it would be interesting to look at the dynamics of this and figure out what affects the formation of that active state of the enzyme. And I'm just going to very briefly tell you about some experiments that were done by several students, Sam Sternberg, Fuguo Jiang, David Taylor

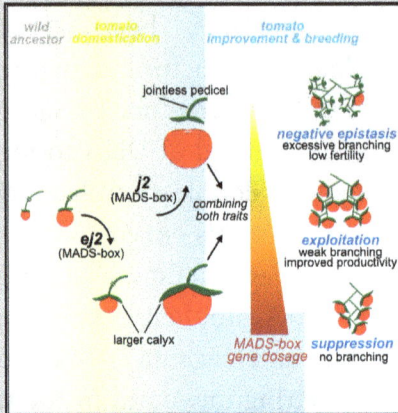

Cell

Article

Bypassing Negative Epistasis on Yield in Tomato Imposed by a Domestication Gene

Graphical Abstract

Authors

Sebastian Soyk, Zachary H. Lemmon, Matan Oved, ..., Dani Zamir, Yuval Eshed, Zachary B. Lippman

Correspondence

lippman@cshl.edu

In Brief

Why did a crop domestication gene hinder breeding with a modern breeding gene responsible for the beneficial "jointless" trait in tomato, and how can this genetic interaction be overcome and exploited?

and then more recently by Yavuz Dagdas and Janice Chen together with a colleague at Berkeley Ahmet Yildiz. And so what we've been doing with Ahmet's lab is putting pairs of dye molecules into Cas9 in places where we think we can detect this conformational rearrangement of the protein using a method called FRET, which stands for fluorescents resonance energy transfer. You can just detect an actual transfer of energy between these dye molecules, but only when they're close enough to interact. And so, by doing this using a single molecule system, we can actually tether Cas9 with its Guide RNA onto the surface of a slide and then flow in DNA molecules, and we can control what sequence of DNA we're using here either using a perfect match to the Guide RNA or including... or using DNA molecules that have mismatches to the Guide RNA. And then, we can ask how those affect the rearrangement of the protein structure by monitoring changes in these FRET interactions

Molecular Therapy
In Vivo Excision of HIV-1 Provirus by saCas9 and Multiplex
Single-Guide RNAs in Animal Models

3 May 2017

Yin et al. use multiplex CRISPR/Cas9 genome editing technology to excise
the HIV-1 provirus in a precise manner in three different HIV-1 animal models
via *in vivo* AAV gene delivery. The feasibility of HIV excision in infected cells
in vivo paves the way toward human clinical trials to cure HIV-1 infection.

between pairs of dyes. So, I'm just going to show you one slide that shows what some of these data look like and what we were able to figure out using this method.

And so, what I'm showing you here, so again this is the way the experiment is set up. So, we have the Cas9, protein with its Guide RNA that's tethered to a slide and we're flowing in different DNA substrate molecules. And then, we're just monitoring the change in fluorescents resonance energy transfer that happens as a function of different kinds of DNA substrates. And what you can see here is that when we use a fully matched DNA substrate, we get most of the molecules occupying a confirmation that we can monitor shown on this FRET scale right here that indicates a fully active Cas9, a Cas9 that can cut the DNA. We can test that and show that that's true biochemically. But what's very interesting is that, as you introduce even one base pair mismatch between the DNA and this 20 nucleotide Guide RNA, you start to see population of a state right

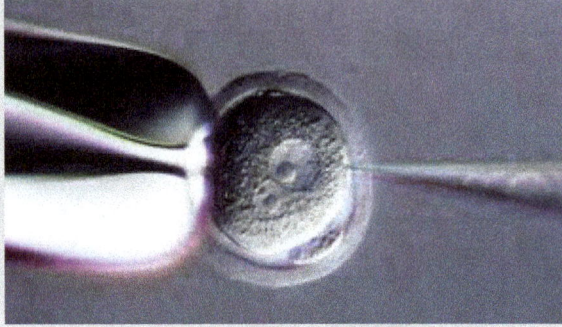

What about human germline editing?

here, shown in green, that is we found out as we compared these in intermediate state. And then over here on the far right-hand side in blue is a state that is fully inactive. So this is the RNA only confirmation of Cas9 over here before it binds to DNA. It's in that fully inactive state. And so, what was really intriguing was that with different numbers of mismatches between the DNA and its Guide RNA, you can observe that many of the molecules get trapped in this intermediate state and that turns out to correspond to this rotation of that HNH domain, the cutting domain, swinging into place to cut the DNA. So, it really only swings into place efficiently when the system is engaged on a fully-matched target sequence. And that's just sort of summarized. Here, where we identified this, what we call a conformational checkpoint that allows the system to basically allows this protein to ask, am I on a fully matched DNA or DNA that should be, maybe it's a virus and I should cut it, or am I engaged on a mismatched sequence in which case I probably don't — this probably shouldn't be cut and that traps it more of the time in this conformational intermediate state that is not active to cut DNA and can more often just associate. That's one of the mechanisms this system uses to ensure accuracy of DNA recognition and cutting.

HUMAN GENOME EDITING
Science, Ethics, and Governance

Committee on Human Gene Editing:
Scientific, Medical, and Ethical Considerations

A Report of
NATIONAL ACADEMY OF SCIENCES *AND*
NATIONAL ACADEMY OF MEDICINE

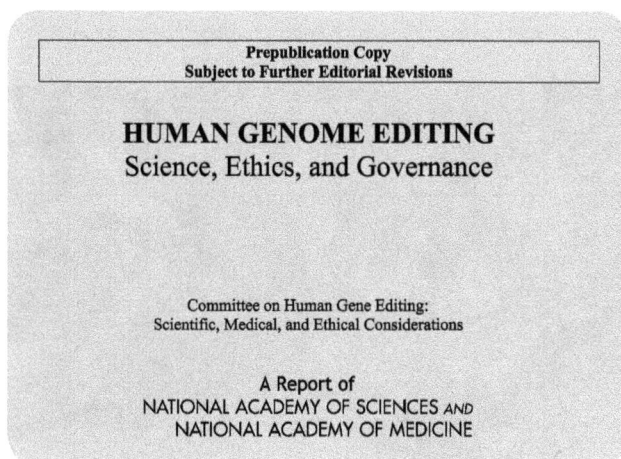

So, it turns out that in nature, you might imagine that you know as the systems are sort of evolving and they are evolving in an environment where there's lots of phage, and so phage are constantly adapting and figuring out ways to get around the CRISPR system that you might find ways that phage fight back. And in fact, very recently, just so in the last year, Alan Davidson's lab at the University of Toronto and now more recently Joe Bondy-Denomy at UCSF have actually figured out that there are proteins that phage make that are called anti-CRISPRs. And this is just one example of these and these were found in a very clever way. And so, work that was done originally by April Pawluk when she was a graduate student in Alan Davidson's lab, what they figured out was that if you looked in bacterial genomes that have incorporated an entire phage sequence that we call a prophage to survive and avoid getting cut by the CRISPR system. If it's acquired a sequence that can recognize a sequence in the prophage, the phage has to encode an anti-CRISPR protein that shuts down the CRISPR system. And that's exactly what happens. And so, April, I have to give her a lot of credit. This was really her idea as a graduate student was to go out looking for these and she actually found examples and now many people are digging up lots more examples and we now know these are very,

Acknowledgments

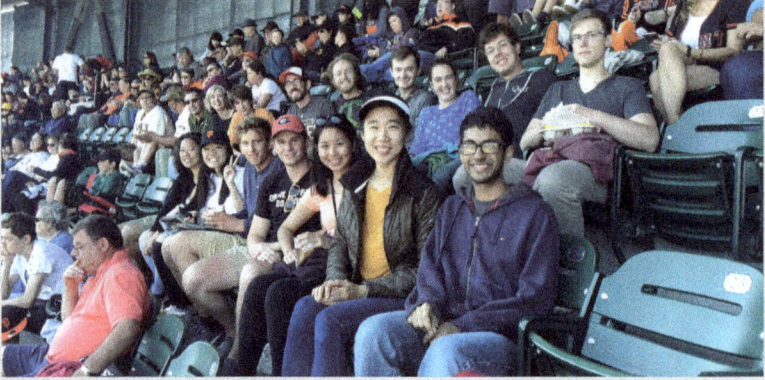

Kryz Chylinski
Emmanuelle Charpentier – MPI Berlin

Yavuz Dagdas, Ahmet Yildiz – UC Berkeley

Karen Maxwell, Alan Jacobson – Univ. of Toronto
Erik Sontheimer – U. Mass Medical Center
Joe Bondy-Denomy – UCSF

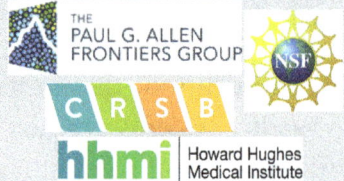

THE
PAUL G. ALLEN
FRONTIERS GROUP

NSF

C R S B

hhmi | Howard Hughes
Medical Institute

very abundant in nature. And this is just showing an example where you have a self-targeting sequence in the CRISPR array that gets made into a CRISPR RNA that would cut this prophage in the bacterial chromosome and kill the bacterium, except that this prophage encodes a protein that is an anti-CRISPR able to shut down the activity of this enzyme and avoid getting cleaved. So, as a biochemistry lab we thought, we'd love to figure out how this actually work.

And so, I want to just again largely for the students here, I just want to show you a few experiments. I hope you'll indulge me and just share with you kind of how I like to do research in my lab working with my students and how we try to do the research as questions and answers, where we ask a question and then we figure out an experiment that we can do to answer the question hopefully. Lots of times we don't answer it, but sometimes we get lucky and we do. And then that often leads to the next question, and then it

Company connections

Co-founder:	Caribou Biosciences
	Editas Medicine
	Intellia Therapeutics
SAB member:	Caribou Biosciences
	Intellia Therapeutics
	eFFECTOR Therapeutics
	Driver
Scientific collaborations:	Pfizer, Inc.
	Roche Biopharma

sort of goes on. And so the first question we asked was, you know, if we compare different examples of these anti-CRISPR proteins, and by the way, these are all very small proteins that are typically less than a 100 amino acids inside, so they're very small. And we started working with Alan Davidson and Karen Maxwell at University of Toronto and a few of the ones at their lab had identified. And, what we found out was that there were some really striking differences in the way these different anti-CRISPR proteins actually function. And that led to the finding that we found one that was sort of unusual in a sense because it really turns out to be a broad spectrum inhibitor of Cas9 enzymes. So, this is showing a phylogeny of these proteins. So, if we look at the relationships of different bacteria, sort of illustrated by this phylogenetic tree, and many of these have Cas9 enzymes. And, if you look at the relationship of these Cas9 proteins to each other, they can be really very divergent. There is a few conserve features, but a lot of them are quite distinct from each other. And so, this is an experiment that was done by Lucas Harrington and Kevin Doxzen, two graduate students in the lab, where they're able to take purified anti-CRISPR proteins that are called C1, C2, and C3

and add them to biochemical reactions that we were conducting using purified Cas9 enzymes. And in this experiment, we're actually using four different purified Cas9 proteins that come from different parts of this phylogenetic tree. And what we found, so the way this experiment works is that we take, this is a DNA molecule and we're separating it by size on a gel system here. So, here's un-cleaved DNA, if we add the Cas9 enzyme with an appropriate Guide RNA, we get DNA cutting. And you can see that for all four of these Cas9 proteins. And then here's the cool thing. So, this anti-CRISPR protein called C1 is a robust inhibitor of all three of these Cas9 proteins, not of this one here at the strep pyogenes Cas9 because it's quite distantly related from these. But these are also proteins that are really quite divergent from each other. And just to show you that that's really an unusual property, these two anti-CRISPR proteins here only inhibit one type of Cas9, this one here and not these others. So that we thought that, that seemed very intriguing and work that was done in our lab and also by Erik Sontheimer's lab showed that you could use these anti-CRISPR proteins in cell-based experiments to inhibit genome editing that was triggered by Cas9 and see the same thing that the C1 protein was a diverse inhibitor of Cas9 homologs whereas the C3 protein could only shut down a single Cas9.

And so, then the question was, how is this working? And so, as biochemist, we could ask well, you know, at what step of the reaction that's catalyzed by Cas9 would this inhibitor be blocking? And so, one of the things that was interesting, so this is a little cartoon of Cas9 with its Guide RNA here in black at the top and here is the target strand and the non-target strand of the DNA at the bottom with the two active sites of the enzyme making a cut. And so, these students, Lucas and Kevin, did a biochemical experiment to ask, first of all, does this inhibitor block DNA cutting? And the answer was, it does. And so again, this is looking at — these are we're looking at DNA molecules and their change in size when they get cleaved by Cas9. So, if we just have Cas9 alone, we get lots of cutting, two different enzymes being used here and here. If we use the inhibitor, at the inhibitor you can see that now we're not getting cutting. This experiment was done in a special way. We were able to label either strand of the DNA so we could figure out, is this strand getting cut or is this strand getting cut. And you might or might not be able to see it, but we saw that. For the most part, DNA cutting was turned off except there was a trace of cutting right here for this strand of the DNA. And that made us think that there might be something interesting going on, that maybe the reason that the DNA wasn't getting cut was...had to do with the chemistry of the reaction because you could say, well maybe it doesn't cut the DNA because it just doesn't bind to the DNA. And so, the test whether it binds to the DNA, these students did an experiment in which they took this Cas9, Guide RNA, DNA complex and added this C1 inhibitor to it. And then they asked to figure out whether the DNA stays bound to the protein. They simply looked at its migration in this kind of a gel system where we can separate molecules by size or even big complexes of molecules. And what we found was that when we have the protein alone, we get DNA cutting, which I showed you before, but once we have this inhibitor, now there's a very large sized collection of sort of conglomeration of molecules, a complex of molecules, that is trapped and we can see that by its slow migration in this gel system. So, this really said that what this anti-CRISPR protein does is to trap Cas9 in an inactive but DNA bound state.

So, how do we, you know, figure out where it's binding? And so, the students were able to make a series of truncated forms of the protein and ask which ones of these truncated proteins are able to bind to this anti-CRISPR. And this is just sort of summarizing the result here, and if you were to compare this in detail what you would see is that the common feature of protein truncations that could bind to the C1 anti-CRISPR was the presence of this HNH catalytic domain that suggested that this domain might be directly binding to the C1 protein, something that we could show by separating the C1 complex bound to the HNH domain. This is actually a chromatography column where we separate things by size. And, we found that only when we have this HNH domain of the Cas9 protein plus the C1 anti-CRISPR, we got migration of a bigger, larger sized interacting complex right here. And students went on, they were able to solve a crystal graphic structure of this and they found out that this anti-CRISPR actually binds to the active site of the HNH domain. So, it's literally blocking the site that in this protein that's necessary for chemistry, necessary to cut this one strand of the DNA. And furthermore, this really explains why it's a broad spectrum inhibitor of Cas9 because this is one of the most conserved parts of these enzymes. Lots of other parts of the protein can change, but the fundamental chemical reaction that these are catalyzing stays the same and it's been maintained over evolutionary time. So isn't it interesting? I thought it was fascinating that this anti-CRISPR protein has figured out how to interact with that site. And furthermore, we think it actually blocks the rotation of this HNH domain into position to cut the DNA. So, it's really an inhibitor that takes advantage of this conformational flexibility of Cas9 that we had determined from these biochemical experiments. And this is just summarizing this and we were curious about whether this inactive DNA bound complex might actually be used in bacteria to harness Cas9, not for DNA cutting but maybe for DNA regulation because it actually traps this protein on DNA at sites that are dictated by the Guide RNA. It's something that we're studying now.

And so, just about done, I just want to mention a couple of things about ethics. And, so in thinking about the way that genome editing is being deployed now around the world for technologies that span agriculture, human health, and various kinds of research, one of the things that's become clear is that, things are…the pace of this research is incredibly fast. And, I just wanted to show you two, especially for the students, two recent examples. So this is a paper that just came out recently where a team at Cold Spring Harbor Lab was able to use the Cas9 system to alter genes in tomatoes that make them less prone to branching and breaking off of these branches of fruit. Something that would have been really hard, maybe impossible, to do by classical breeding methods, but now can be done using precision gene editing. And then also, paper that got a lot of attention recently was a team in China that used Cas9 to remove integrated HIV sequences, the virus that causes AIDS from genomes of mice using a very efficient way of delivering Cas9. And, so these are applications, like this example is an application that would apply to an adult animal or if you were doing it in a patient, an adult patient, but there's been also a lot of talk about germline editing, meaning being able to inject editing molecules into fertilized eggs or embryos that would lead to heritable changes to DNA. And people have been doing this from very early days with gene editing systems including Cas9 in animals, but this can also be done in principle in humans. It's certainly been done in primates and that has led to my involvement in a really interesting series of meetings and discussions around this topic and how we encourage responsible use of this technology moving forward.

And, I just want to point out that the National Academies of Sciences and Medicine in the US recently put out this report on human genome editing that really encouraged research using this kind of technology, even in embryos, but suggested a worldwide pause or one might say moratorium on use clinically in humans until it can thoroughly be evaluated as a technology, and certainly from an ethical perspective as well.

So, I think for the students, this kind of just illustrates some of the challenges that happen when you're doing research in your lab. It leads to technology that is going in directions that one couldn't have predicted initially and where, I think it's at that point really important for scientists, all of us, to engage in the conversation about responsible progress. And I'd just like to close by acknowledging a fantastic group of people. This is a picture of folks in my laboratory. We've had, of course, incredible collaborators over the years, Emmanuelle and Christoph and then I mentioned these folks as well and then our funding over here. And thank you for your attention.

The Universe, a Personal View on Exploring the Boundaries, or: The Science of the Extremes*

Gerard 't Hooft[1]

What are the fundamental equations of motion when we consider objects that move faster, have tinier structures, higher mass densities, higher or lower temperatures than ever studied before? Here, we emphasize the importance of studying extreme conditions and situations in science so as to uncover the basic laws of nature. One of the important extremes is extreme complexity, a necessary condition for the emergence of life.

Keywords: Extremely Fast; Extremely Tiny; Extremely Dense; Extremely Heavy; Relativity Quantum; Space-Time; Gravitation.

---

* This chapter was based on a lecture delivered by Professor 't Hooft at the Molecular Frontiers Symposium in the Royal Swedish Academy of Sciences in May 2013.
[1] Nobel Laureate in Physics (1999). Institute for Theoretical Physics, Post box 80.089, 3508 TB Utrecht. The Netherlands.

About Gerard 't Hooft

Gerardus (Gerard) 't Hooft (born July 5, 1946) is a Dutch theoretical physicist and professor at Utrecht University, the Netherlands. He shared the 1999 Nobel Prize in Physics with his thesis advisor Martinus J. G. Veltman "for elucidating the quantum structure of electroweak interactions".

His work concentrates on gauge theory, black holes, quantum gravity and fundamental aspects of quantum mechanics. His contributions to physics include a proof that gauge theories are renormalizable, dimensional regularization and the holographic principle.

After obtaining his doctorate 't Hooft went to CERN in Geneva, where he had a fellowship. He further refined his methods for Yang–Mills theories with Veltman (who went back to Geneva). In this time he became interested in the possibility that the strong interaction could be described as a massless Yang–Mills theory, i.e. one of a type that he had just proved to be renormalizable and hence be susceptible to detailed calculation and comparison with experiment.

According to 't Hooft's calculations, this type of theory possessed just the right kind of scaling properties (asymptotic freedom) that this theory should have in order to yield physically acceptable behaviour at ultra-short distance scales. This was contrary to popular perception of Yang–Mills theories at the time, that like gravitation and electrodynamics, their intensity should decrease with increasing distance between the interacting particles; such conventional behaviour with distance leads to internal inconsistencies, while there were experimental indications that the deep-inelastic domain of field theories tends to be very regular.

When 't Hooft mentioned his results at a small conference at Marseilles in 1972, Kurt Symanzik urged him to publish this result; but 't Hooft did not, and the result was eventually rediscovered and published by Hugh David Politzer, David Gross, and Frank Wilczek in 1973, which led to their earning the 2004 Nobel Prize in Physics.

In 1974, 't Hooft returned to Utrecht where he became assistant professor. In 1976, he was invited for a guest position at Stanford and a position at Harvard as Morris Loeb lecturer. His eldest daughter, Saskia Anne, was born in Boston, while his second daughter, Ellen Marga, was born in 1978 after he returned to Utrecht, where he was made full professor. In 2007 't Hooft became editor-in-chief for *Foundations of Physics*, holding the position until 2016.

On July 1, 2011 he was appointed Distinguished Professor by Utrecht University.

Before being awarded the 1999 Nobel Prize, 't Hooft's work had already been recognized by other notable awards. In 1981, he was awarded the Wolf Prize, possibly the most prestigious prize in physics after the Nobel prize. Five years later he received the Lorentz Medal, awarded every four years in recognition of the most important contributions in theoretical physics. In 1995, he was one of the first recipients of the Spinozapremie, the highest award available to scientists in the Netherlands. In the same year he was also honoured with a Franklin Medal.

Since his Nobel Prize, 't Hooft has received a slew of awards, honorary doctorates and honorary professorships. He was knighted commander in the Order of the Netherlands Lion, and officer in the French Legion of Honor. The asteroid 9491 Thooft has been named in his honor, and he has written a constitution for its future inhabitants.

He is a member of the Royal Netherlands Academy of Arts and Sciences (KNAW) since 1982, where he was made academy professor in 2003. He is also a foreign member of many other science academies, including the French Académie des Sciences, the American National Academy of Sciences and American Academy of Arts and Sciences and the Britain and Ireland based Institute of Physics.

This account is a thorough rewriting of a recording of a talk I gave with this title in the Swedish Academy of Sciences, where I happen to have left many pleasant memories. Most other talks I attended at this meeting, were about life sciences, which, due to their severe complexity, are difficult to fathom by a pure sang theoretical physicist such as myself. It so happens however, that some of our interests do coincide. One question sometimes posed by biologists concerns the origin of left-right asymmetry in living organisms. Actually, the history of the evolution of life forms on planet Earth shows that living organisms began by having much more symmetry: they may have started out being practically spherical. Then, when organisms needed a head and a tail, spherical symmetry was reduced to cylindrical symmetry. Subsequently, living organisms such as snails, discovered that structures become sturdier when you curl them up. These organisms, such as the ammonites, were left with only left-right symmetry. Finally, most modern snails also break that symmetry, which again seems to have improved the sturdiness of the structure of their shells.

Here, the question is whether and how a competition between life forms and their mirror images may have emerged, with the predictable outcome that one of the two won. Most snails seem to be twisted in the same "right-handed" direction. Why not left-handed? Whence this asymmetry? Some researchers suspect that the origin of this asymmetry can be traced back to the minute mirror asymmetries that were found to exist in the fundamental laws of Nature. For physicists such as me, however, a simpler answer to the asymmetry question seems to be obvious. All these symmetries probably broke *spontaneously*, as we often see happen even in non-living materials, such as magnetic material, and even sub-atomic particles. No external asymmetric force is needed at all for symmetries to get broken. Note that our environment breaks left-right symmetry in a big way, an asymmetry that cannot be directly attributed to asymmetries in the fundamental laws of physics.

For the life sciences, life in extreme conditions is often interesting. How does life endure extreme cold, extreme heat, extreme acidity, the extremely high pressures of the deepest trenches in the oceans, and so on? As a scientist, I am particularly interested in the conditions on the other planets and asteroids in our solar system. Being extreme in many ways, they form interesting challenges for human life when we contemplate the possibility of establishing human colonies there. The conceivable possibility to establish a human colony on planet Mars is seriously being considered today, not only by agencies such as NASA, but also by private groups, such as "Mars One", established by some compatriots of mine. This however can better be left for an other talk.

From a physics point of view, the conditions just mentioned are not extreme at all. Even the center of the Earth, and the Sun, are not yet overly extreme from a physical point of view. We know of planets, stars, and regions in the distant or the early universe, where conditions are/were far more extreme. My main subject today is the extremes in physics. We can extrapolate the laws of non-living matter very well in most places, but there are limits where also our science becomes more exciting, more questionable, and eventually incomprehensible. What physical science and life science have in common is, that studying the extremes is very instructive for learning what the general laws are. So we do experiments in, and theoretically analyze, the extremes.

Let me consider various possible extremes one by one, and disclose what was discovered about our physical world when subject to these extreme conditions.

Extremely Fast

According to the old laws of physics, any physical structure, regardless whether it is living or dead material, can be moved around with velocities that, in principle, can be arbitrarily large. This is not as evident as it may sound, since we can imagine universes where such would not be the case — if you would try to move something, it would immediately break into pieces.

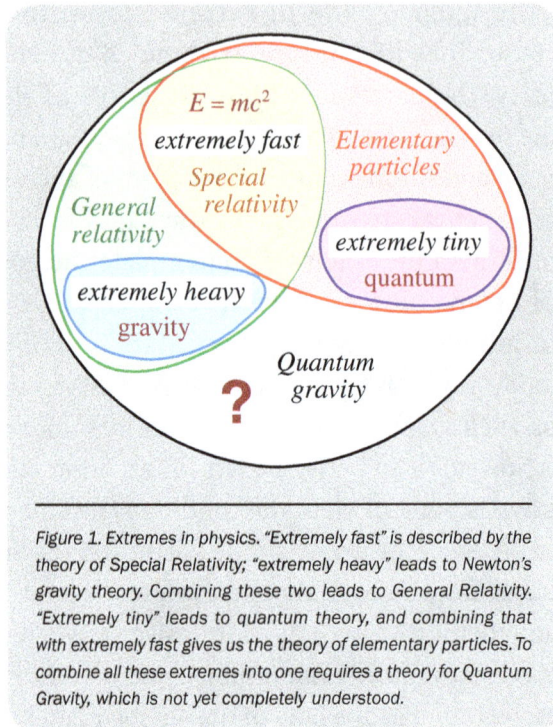

Figure 1. Extremes in physics. "Extremely fast" is described by the theory of Special Relativity; "extremely heavy" leads to Newton's gravity theory. Combining these two leads to General Relativity. "Extremely tiny" leads to quantum theory, and combining that with extremely fast gives us the theory of elementary particles. To combine all these extremes into one requires a theory for Quantum Gravity, which is not yet completely understood.

It was Albert Einstein who asked the question what an observer would see if (s)he moves as fast as the speed of light; when observing the surrounding universe, using light to observe things, you might, for instance, see nothing when you look behind you. Einstein's extremely ingenious discovery was, that one can imagine slight modifications of the laws of physics known at his time, in such a way that an observer would be able to look around and use light to observe things, exactly as if he were not moving at all, except for one limitation: (s)he cannot actually reach the speed of light regardless his/her efforts.

The speed of light is extremely large. In fractions of a microsecond light can move by hundreds of meters. This is why, in the older text books of physics, observers moving that fast were simply not considered. But since, today, most of our communications go with the speed of light, and contain not millions but billions of data per second, this domain that used to be extreme, has become very central in physics, ever since the beginning of the 20th century.

The space we live in is three-dimensional. To locate any object, any point, in space, one needs exactly three coordinates, usually called x, y and z. It was Hermann Minkowski who proposed to add time, t, as a new dimension, so that three dimensional space was replaced by a four dimensional space-time. Think of a train schedule, where it is indicated where the train is (x and y together localize the train on the map, z indicates how high it is above sea level), but also we add at what time, t, the train is supposed to be there.

When you change the velocity of something, this can be regarded as a rotation in Minkowski's space-time. In three-space, rotations are easy to visualize, and a physical law says that the internal properties of something do not change when you rotate it. This is a fundamental property of space. At first sight it seems not to be exactly valid: if I rotate a glass of water along a horizontal axis, the water may pour out. That, however, can be attributed to the presence of planet Earth below us, which possesses a gravitational field attracting the water. Similarly, changing the velocity of something can be regarded as a rotation in Minkowski space-time, and the fact that the internal properties of something, such as the question whether it is dead or alive, sick or healthy, do not change when you make such a rotation, is a fundamental property of Minkowski's space-time. Again, if the property seems to be not exactly valid, for instance if you feel more and more wind when you increase your speed, this can be attributed to the presence of air, which has a fixed velocity itself.

There are two different ways in which we can consider rotations: you can either rotate the object you are looking at, and compare what you see under the different conditions, or you can rotate the observer, that is, yourself, and look at the same object. In both cases you can ask whether the laws of physics themselves are affected by these rotations. In both cases, you modify the coordinates x, y, z, and t, by the rotated coordinates. Einstein found how one can formulate the laws such that they are guaranteed not to be affected by rotations, be it in three-space, or in Minkowski space-time.

The way things can rotate in space and time had already been formulated by my compatriot Hendrik Antoon Lorentz, in 1892 and 1895, well before Einstein (1905). There is something odd, however, about these rotations. Why is it that the speed of light is a limiting speed? The answer to this is that the rotations in Minkowski space-time are non Euclidean, so they differ from ordinary rotations. While in ordinary rotations the distance between two points always remains the same, rotations in Minkowski space are chosen to be such that the speed of light is always kept to be the same regardless by how much we rotate the coordinates.

A point on an object that rotates in ordinary space describes a circle. It can rotate around the circle as many times as one wishes to consider. If I would use different units of length for horizontal distances than for vertical distances, a point on an object that rotates along a horizontal axis would describe an ellipse — a flattened circle. Circles and ellipses are conic sections, but there exist also other conic sections: the parabola and the hyperbola. In Minkowski's space-time, a point on a rotating object may follow a hyperbola.

A hyperbola has two asymptotes; these are straight lines that are approached as closely as one wants by the hyperbola, but they are never touched. A wave of light, or a photon, follows the path on the asymptote; a rotated object cannot cross that asymptote, and this is why ordinary objects can never reach the speed of light — no matter how far one rotates it, the point has to stay on the hyperbola.

Lorentz had discovered that these are the rotations that leave the laws of electricity and magnetism unchanged. It was Einstein who made the next step: *all* laws of physics must stay unchanged when you consider these rotations, now called Lorentz rotations. And this does imply that Newton's laws of mechanics, the ones that were extremely successful in explaining the motion of planets and moons in elliptical orbits, had to be modified. At first sight, the modifications seemed to be minute, but the consequences for our physical world were gigantic.

In particular, it was found that whenever energy is added to a system, its mass will change: the mass change was found to be equal to the energy divided by the square of the speed of light, c^2. This change is minute, simply because c^2 is so big.

But it was already known that the atomic weights of the elements deviate slightly from what one would have expected if these atoms were composed of elementary particles, and these mass differences could be measured precisely. Now, Einstein realized that the mass changes were explained in his theory; they correspond to the energy by which the fundamental particles are bound together in the heavier atoms. If we call the observed mass defects m, then the associated energies must obey $E = mc^2$. It was thus deduced that the nuclei of the atoms of the elements in the periodic table contain huge amounts of energy. The astounding consequences of this observation changed the world.

In ordinary space, a particle is just a point. In Minkowsi space-time, we see the particles somewhere at every moment t in time, and therefore, a particle is represented by a curve instead of a point.

At every given time t, the points in three-space form a section of Minkowski space-time, called the Cauchy surface. When we perform a Lorentz rotation, this Cauchy surface also rotates, it does not stay the same. This means that the question whether two events in Minkowski space are *simultaneous*, will be answered differently by observers with different velocities. Thus, the concept of simultaneity is relative, not absolute. This gives rise to many counter intuitive features in Einstein's relativity theory.

Since now we know that objects with any material substance in them, cannot surpass the velocity of light, we can now slightly modify our conception of extreme velocities: Velocities are extreme when they are extremely close to the speed of light. To reach such extreme velocities, extremely large amounts of kinetic energy have

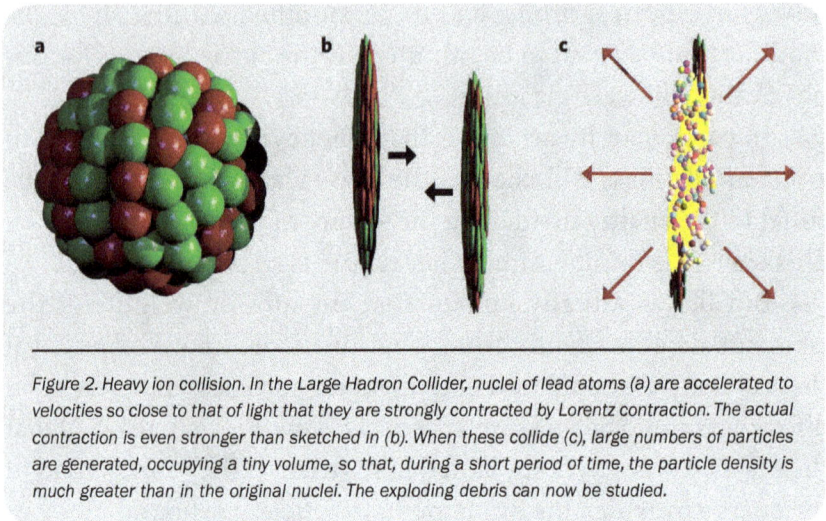

Figure 2. Heavy ion collision. In the Large Hadron Collider, nuclei of lead atoms (a) are accelerated to velocities so close to that of light that they are strongly contracted by Lorentz contraction. The actual contraction is even stronger than sketched in (b). When these collide (c), large numbers of particles are generated, occupying a tiny volume, so that, during a short period of time, the particle density is much greater than in the original nuclei. The exploding debris can now be studied.

to be imparted to the object. What happens when such extremely energetic objects collide against one another has become a separate branch of physics: high-energy physics. The Large Hadron Collider, a particle physics laboratory near Geneva, Switzerland, is today's champion in reaching the most extreme velocities in this sense. Hadronic matter is accelerated there to more than 99.999 999% of the speed of light. As of what happens at velocities even closer to the speed of light, there are still mysteries.

The theoretical doctrine needed to understand the extreme velocities is Einstein's theory of special relativity. There is however a more generalized relativity theory, General Relativity. About that, more later.

Extremely Tiny

Living organisms such as bacteria and viruses are tiny, but not at all as tiny as the tiniest structures physicists are studying. The discovery that extremely tiny objects (the sub-microscopic world) obey laws of physics that are fundamentally different from those of the macroscopic world (people, cars, planets) was made also at the beginning of the 20th century. At that time, the strongest evidence for the existence of atoms and molecules came from the study of the laws of chemistry. In many respects, atoms and molecules behave very differently from larger and heavier objects, such as bacteria, let alone people and planets.

First of all, when two atoms are equal, they really are fundamentally equal. Any observable differences in behavior are irreproducible, or accountable to environment or other factors not under our control. When an experiment with beams of atoms or other fundamental particles is carried out, then, apart from purely statistical fluctuations, atoms of the same species are indistinguishable, and the outcomes of statistical experiments are completely reproducible. The doctrine required for the description of these objects is very different form classical mechanics, and it is summarized by just one word: Quantum.

Differently from relativity theory, Quantum Mechanics was developed by the basic insights of many scientists, such as Niels Bohr, Louis de Broglie, Werner Heisenberg, and Erwin Schrödinger among others. It also took much more time to understand it as we do now, and even now, it is much more difficult to explain.

The first thing physicists learned is that very tiny particles are associated to fields that form wave patterns. These waves contain all the information needed to describe a particle. Different species of particles carry — or are carried by — different kinds of oscillating fields. The frequencies of the oscillations are indicative for the energies these particles contain, while the wavelengths are linked to momentum, that is, velocity times mass. These observations may sound somewhat vague, but it was discovered by Erwin Schrödinger that the waves obey mathematical equations, wave equations, which turn out to replace the old Newtonian laws of mechanics: quantum mechanics. With these equations, quantum mechanics became a theory with high precision.

The amplitudes of the waves determine where the particles can most likely be found, and again these features were precisely laid down by mathematical equations. The equations tell us that the waves form certain patterns, and the ways these patterns move around tell us how particles move. It turns out that, if many, heavy particles are put together, the equations are such that they can be replaced by the older Newton equations, and these classical equations are much more accessible to intuitive understanding.

These discoveries came hand-in-hand with a much better understanding of the properties of atoms, beginning with Niels Bohr's crude model of the atom, and culminating into precision physics that allowed us to check and understand atoms fantastically precisely.

An atom consists of a very tiny nucleus, composed by what we call hadronic material, surrounded by electrons that orbit the nucleus at relatively larger distances. These electrons form wave patterns that become more and more complex as we consider larger and heavier atoms. The complexity comes from the fact that larger atoms contain many electrons. Carrying just one electron, the hydrogen atom is mostly spherically symmetric, but the 8 electrons orbiting oxygen nuclei form waves with four peaks, roughly taking the shape of a tetrahedron. When oxygen binds to hydrogen, the hydrogen nuclei are attracted by these peaks, and consequently, when two hydrogen atoms are bound to an oxygen atom, they do not sit opposite to each other, but form an angle of 105°. We are talking of a water molecule here, and its special shape gives water its unique properties. The description given here is very crude — the actual angular separation between two points of a tetrahedron is 109.5°, which makes clear that the tetrahedron does not explain everything; the behavior of the 10 electrons in a water molecule is very complex.

The tendency of carbon atoms to arrange themselves in long chains can be understood along similar lines. Quite generally, due to their wave patterns, electrons in atoms force other atoms to be attached under different angles, which give the resulting molecules their special shapes. These discoveries brought us closer to an understanding of the complex laws of chemistry that lie at the basis of life.

We reiterate that, by studying the extremes, physicists discover how laws of Nature may have to be reformulated to cope with the new situations. This will also be true for the other extremes that are to be discussed later.

The physics of atoms is not at all at the limit of the very tiniest things we can study. The atomic nucleus is much smaller than the atom itself, and it is controlled by its own physical laws. These laws were at the center of theoretical and experimental physics around the mid 20th century. By that time, it was understood that nuclei consist of protons and neutrons, particles of nearly equal mass, but totally different in their electromagnetic properties. The nucleus does not consist of protons and *electrons*, which had been the natural thing to suspect in earlier days. Today we know that the quantum mechanical laws controlling the electron do not allow it to occupy such small spaces as what we have inside a nucleus. The electron is far too light for that.

Protons and neutrons exert quite strong and complex forces on one another. By 1980, these forces were quite well understood qualitatively, but several decades were needed before accurate agreement was found between experimental measurements and theoretical calculations. We have even smaller building blocks, the quarks, of which there are 6 basic types. Three of these quarks are needed to form a hadronic particle such as the proton and the neutron. In addition, there are relatives of the electron, called leptons, of which we have also 6 types all taken together.

Extremely Tiny and Extremely Fast

To understand the laws of physics at scales as tiny as the atomic nucleus, we have to realize that all structures at such scales are generated by elementary particles having great velocities. This is the way it must be according to the laws of quantum mechanics. Thus, the physics of very tiny things on the one hand, and the laws of the greatest velocities on the other, merge when the extreme limits are concerned. And so it happens that the Large Hadron Collider, in its attempt to study the tiniest distance scales in physics, is designed also to produce particles with the highest possible velocities, thus combining two extremes in one.

Clearly, this limit now combines Einstein's special relativity theory with quantum mechanics. How to do this "quantum relativity" right, was a major problem during the 20th century, but it was almost completely solved around the 1970s. The solution is called Quantum Field Theory. Many physicists contributed to this success. Paul Dirac discovered a relativistic equation for describing the wave function of an electron with relativistic velocities. He discovered that the equation works best if one allows for the presence of unlimited numbers of electrons, in addition to something completely new: positrons, which are the antiparticles of electrons. It was essential for this theory to work, that under certain conditions one can observe the successive creation of a particle and its annihilation shortly afterwards, but that this process becomes indistinguishable from the process where first an antiparticle is produced, after which that is annihilated again, where the points of creation and annihilation are interchanged.

Because of the quantum nature of these particles, one can sometimes not exactly tell whether its velocity was slower or faster than light, and even whether the particle went backwards in time. The crucial thing to keep in mind here was that the elementary field equations that describe all this, should obey the principle of causality: never allow any piece of information to be sent faster than light, but don't worry whether it was carried by a particle or an antiparticle.

The theory that ensued was based on the notion of fields, to which quantum equations were imposed, hence the name quantum field theory. It eventually evolved into a rigorous scheme, forcing the behavior of all known elementary particles to be described by a concise set of equations: the Standard Model. All known particles now carry three types of charges: first we have the electromagnetic charges, which were already known and understood since James Clark Maxwell wrote down his famous equations around 1861. Since Dirac's work, we know how the electromagnetic force acts upon the tiniest elementary particles. Secondly, we have the weak charges. The weak force caused by these charges actually mixes with the electromagnetic force, which is why today we speak of the electroweak forces.

Finally, there is the strong force. This force only acts on what we call *hadronic* particles. In contrast to the electric charges, which are simply either positive or negative, the strong charges come in three types, for which we use the analogy with color: red, green and blue. The antiquarks have the conjugated colored charges: cyan, magenta and yellow. These forces are so strong that their action can be very non-linear, which makes the calculations concerning the strong force extremely complicated: during the short period of a collision of particles, the strong force can act many times. The forces by which these charges act on one another are a generalization of the concept of electromagnetic force, called the Yang-Mills force. While electromagnetic forces are carried along by a single type of particle, the photon, the other forces are carried by three kinds of weak photons, called W^+, W^- and Z°, and eight types of strong photons, the so-called gluons. While the photon was a neutral particle, the gluons do carry color charges themselves.

There is one very special force in nature that is not, or not yet, captured by our understanding of the Standard Model, which is the gravitational force. Under normal circumstances, the gravitational force is so weak that it hardly affects our understanding of the physical laws concerning particles of matter. The hadronic particles circulating in the Large Hadron Collider do not feel Earth's gravity simply because they are moving much too fast for that.

However, when we penetrate the world of the tiny things deeper, we need to look at particles with ever increasing amounts of energy, because that makes their waves tinier. Gravity couples to mass, and mass is energy, as Einstein had discovered, and this is why the gravitational force slowly but steadily gains in strength when we continue our journey into the things of the small. Eventually, gravity will overtake all other kinds of forces. When this happens, we approach a new extreme limit. We shall return to that.

Extremely Heavy

When objects possess a large amount of mass, they tend to become heavy. This is a direct consequence of the fact that the gravitational forces only depend on mass and nothing else. When we study heavy objects, we are eventually led to heavy objects in outer space: planets and stars. These objects are sensitive to their own gravitational fields, and as soon as they reach a certain limiting amount of mass, they attain the only allowable stable shape, that of a sphere — possibly flattened to some extent because of rotation.

These heavy objects are surrounded by a gravitational field, which accelerate other, nearby objects. As long as these gravitational fields are not too strong, they present little problems when we wish to describe them in terms of physical equations. As was discovered by Johannes Kepler, and explained by Isaac Newton, the orbits of stars and planets in the gravitational fields of others, take the shape of ellipses.

Extremely Dense

When the gravitational field reaches certain thresholds, particles inside a gravitating body may undergo phase transitions into forms of matter we still know fairly little about. Such a thing happens in a "neutron star". Here, the original atoms are compressed to the extent that protons and neutrons hardly fit in the space they have. Most of the electrons do not fit there at all and they are forced to move out. This in turn causes most of the protons to absorb electrons and turn into neutrons. Thus, mainly neutrons remain inside. This results into what we call a neutron star. Neutron stars are observed because they tend to rotate very fast, hundreds of rotations per second. All radio emissions fluctuate in accordance, making them easy to recognize as pulsating sources of radio signals, or *pulsars.*

To illustrate how extreme these densities are, we note that our sun would turn into a neutron star only if we could compress it such that its diameter is a few tens of kilometers. The actual diameter of the sun is over a million kilometers.

Other features also tend to become extreme in the insides of a neutron star. Its magnetic field, for instance can become hundreds of millions times stronger than fields produced in our laboratories. If a pulsar would approach us as close as where the moon is now, our watches would stop running. The emitted radiation would be lethal for us at that distance. One pulsar has a rotation rate clocked to be over 700 revolutions per second.

Even higher densities can be considered. In the Large Hadron Collider, nuclei of lead atoms, with all 82 electrons stripped off, can be made to collide against one another. During the collision, for a short time interval, the protons and neutrons of these nuclei are forced into a space that is even smaller than inside a neutron star. Also the temperatures are the highest that can be obtained in a laboratory on Earth.

During the earliest phases of the universe's existence, the universe expanded very rapidly, and it all began when all particles constituting the presently observable part of the universe, were compressed into a single point. Up to times less than a microsecond after the beginning, the conditions were much more extreme than in any laboratory on Earth. Also the temperature was much higher than what we can study in our machines. To some extent, one can calculate how matter must have behaved, but when densities and temperatures were much higher, phenomena occurred that we can only speculate about. However, we know what our universe looks like today, and from that we can infer at least some properties of matter during the initial phases, since these must have been the reason why the universe expanded very rapidly and homogeneously to become what it is now. We shortly return to the beginning of the universe later.

Extremely Heavy and Extremely Fast

Once again, all changes when, due to the action of very strong gravitational fields, objects moving near it, and sometimes also the material out of which the source object is made itself, attain velocities nearing that of light. What we then see is relativistic gravity. Here, mass and energy get intertwined, and the energy residing in the gravitational field itself can no longer be ignored. Very precise measurements in ingenuous experiments have revealed that gravity always acts on mass and nothing else. This gives us severe restrictions when designing a theory for relativistic gravity. The solution was found by Albert Einstein in 1915: General Relativity. His theory combines extremely heavy objects with extremely high velocities.

To understand what happens, our picture of space and time has to be modified once more. Since high velocities amount to rotations in Minkowski space-time, one finds that such rotations are generated near a heavy gravitating body. This causes all sorts of deformations in space-time, and the net result of this was that Einstein was forced to contemplate *curvature* in space and time.

The most conspicuous effect of this curvature is due to the curvature in the time direction: inside the gravitational potential well of a gravitating body not only all clocks seem to proceed slower than at farther distances, but also all other physical processes appear to slow down. A local observer will not directly notice this, because his own body and all his measuring devices slow down as well, so we are talking of a property of space and time themselves.

Also in the purely space-like direction one finds curvature to occur. Distances between points in space deep inside a gravitational potential, appear to be greater than if we used a flat coordinate frame and counted the lines in between.

It all becomes a coherent theory only if we add the equations, but these tend to be rather complicated, since curvature depends on direction, in a very intricate way. It so happened that Einstein did not have much choice. Only one theory was possible, with the exception of slight, unknown modifications, possible only at the extremely large scale given by the size of the universe itself.

Thus, the solution as to what happens in this extreme case was found, and it was basically unique. General relativity came with various possible ways to test it. Direct conflicts with observations were never encountered, although we do notice gravitational fields inside and around galaxies that are difficult to account for, if they are assumed to be due to matter at all. The missing source of these gravitational fields is called "dark matter", but the possibility that General Relativity itself requires amendments is seriously studied by quite a few researchers.

"Extremely heavy and extremely fast" has an inescapable limit. General Relativity allows us to compute precisely what happens when the gravitational force of a heavy object surpasses a magic boundary point: particles can be gravitated towards the speed of light, and at first sight nothing may seem to stop them from continuing their acceleration beyond that. What happens then, however, is that space-time curvature reaches a turning point. The equations are quite clear about this: the gravitating body then collapses entirely, and it turns into what is known by the name "black hole".

Since its equations are known and understood, we know quite a lot about black holes. They are extremely violent. They are surrounded by a mathematically defined surface called the *horizon*. In principle, one can ask what an observer will see who falls through this horizon. Besides experiencing queer gravitational forces acting on his body, the observer can, in principle, survive the journey through the horizon, but very quickly this observer will end up in a kind of apocalypse that no-one can survive. Only if the original black hole was deformed by rotation, there may still be another universe waiting ahead of our observer, but contact with our own universe will have terminated.

Black holes are black, nothing will come out of them, unless we combine the extremely heavy and the extremely fast, with the physics of the extremely *tiny* distances. There, quantum mechanics tells us a different story.

Extremely Heavy, Extremely Fast, and Extremely Tiny

This is the one extreme limit where not only strange things happen, but we are really extremely ignorant about the exact way to describe it. What we are fairly certain about, is a theoretical discovery made by Stephen Hawking in the 1970s. He found that a new prediction can be deduced by applying the laws of quantized fields to the curved space and time of a black hole. Even if an observer falling into a black hole would describe his environment as being completely void of any elementary particles, the outside observer will detect particles coming out. The reason for this is that the outside observer experiences *time* in a way that is fundamentally different from what observers see who fall in: for the outside observer, time at the horizon will seem to come to a halt, while someone going in will observe that time continues. Time is mathematically linked to the notion of energy, and energy is used to define what a particle really is: an object carrying a certain amount of energy. Consequently, observers outside a black hole will observe particles even where observers falling in see none.

Important as this discovery is, it does not exactly yield the complete set of equations needed for a precise description of the world in this extreme limit. Candidate theories that might do the job have been proposed. Some theories posit that gravitational forces might eventually become independent of scale, so that, in this limit, we simply approach an asymptote, where no more things of interest will happen; this theory is called "gravitational safety". Another theory takes it that space and time can be partitioned into small domains, connected by lines and closed loops; this theory is called "loop quantum gravity". The most powerful and imaginative approach is collectively called "superstring theory". This theory comes in various versions, such as "M-theory", where the strings are replaced by giant matrices. However, the promises of all these theories are often exaggerated by their proponents. Fact is that, we still do not know or understand the resulting equations very well.

Extremely Tiny and Extremely Big

There are other interesting extremes. We may consider very big quantities of very tiny particles. This could enable us to consider theories for the entire universe. The universe is old, some 13.8 billion years. In the beginning of the 20th century it was found, by Edwin Hubble, that the universe is rapidly expanding. This means that, in the far past, the universe must have been a lot smaller than at present, and all evidence is pointing towards the idea that the entire part of the universe that can presently be seen in our telescopes, may have originated in a single point.

The history of the universe has been reconstructed precisely in recent years. Saul Perlmutter, Brian Schmidt and Adam Riess received the 2011 Nobel Prize for their work in determining the expansion rate of the universe. For this they had used observations of supernova explosions of stars. A supernova can also be considered as the most extreme events that can take place inside a star, destroying most of it. Supernovae can be used as standard candles to shine more light on the exact structure the universe took from the very first times onwards. Surprisingly, they found that the universe expands by an accelerating pace. Before this acceleration effect was known, there were some discrepancies in the determination of the universe's age. Now, these discrepancies have been removed, but the acceleration was left to be explained. Most likely, the cause is a term in Einstein's General Relativity theory that is not very well understood: the cosmological constant. This constant was usually assumed to be zero, but now it is realized that this cannot be so. This constant is one of the smallest non-vanishing constants we have in describing the physical world, and we do not know why it is there.

As stated before, in earlier epochs, the universe must have gone through a phase where it was extremely dense. It must also have been extremely hot, since expansion goes with cooling down. Careful observations and theoretical deduction have revealed that the initial expansion rate must have been enormous; the universe must have been expanding in an exponential fashion. This would be needed if indeed everything originated in a point, which in turn is a much favored picture just because the early universe was observed to have been extremely homogeneous. Even the most distant extents of the universe must all share a common ancestor point in their pasts.

We have the equations required for an exponentially expanding universe. This "inflationary universe" results if we assume that matter in the distant past were in a more symmetric phase than what it is in at present. The theory agrees with the minute fluctuations observed in radiation originating from these early phases our universe has been in. This subject has presently evolved into an independent branch of science, called cosmology. Less than half a century ago, speculations on this topic could be dismissed as "science fiction". Today, it still contains fiction, but the scientific part of this doctrine has matured immensely.

Extremely Cold

At low temperatures, particles of matter tend to move very slowly. While at ordinary temperatures most properties of matter can be understood using ordinary, Newtonian mechanics, very low temperatures require quantum mechanics to understand what happens. These features can indeed be exotic. At the beginning of the 20th century, Dutch physicist Heike Kamerlingh Onnes decided to study the properties of atoms by investigating metals at the lowest temperature he could achieve, which was of the order of one degree above absolute zero. When checking the electric conductance of mercury, which is frozen solid at these temperatures, he found something totally unexpected: all resistance against electric currents disappears, or, mercury becomes superconducting.

At ordinary pressures, all substances except helium become solid at sufficiently low temperatures. Besides mercury, many other materials were found to become superconducting. This is a quantum effect, only understood much later. In helium, the isotope ^{4}He becomes a superfluid: all resistance against flow disappears. When this material is pumped through a pipe, it shows no resistance at all (as long as the velocities are not to great). This is also a quantum mechanical effect, similar to superconductivity. Kamerlingh Onnes had encountered superfluidity of helium as well, as was later deduced from the notes he made, but he had not interpreted what he saw the way we do now.

Today, there still is a lot of interest in low temperature physics. Individual atoms, forced to sit in certain patterns, are now cooled to less than a millionth of one degree Kelvin. Many exotic phenomena take place in these extreme circumstances.

Extremely Complex

Finally, our universe is extremely big, consisting of gigantic amounts of particles, and consequently these particles must be behaving together in a highly complex manner. Needless to say, only extremely complex phenomena can be held responsible for many features occurring in our part of the universe. Without doubt, the most complex feature of all will have been the emergence and evolution of life on at least one planet. So we return to the life sciences, where this talk also began.

Only recently a long held suspicion has been confirmed: there exist enormous numbers of planets near the stars surrounding our Sun. Many of these planets are very different from planet Earth, but also gigantic numbers of planets must exist that are very similar to Earth. Are these also teeming with life, or is our planet a single, lonely exception having living creatures accidently colonizing it, changing its surface completely?

Complexity raises several other mysteries concerning our universe. As the reader may have noticed, combining several extremes together leads to situations that are even more difficult to comprehend. When combining extremely heavy with extremely fast and extremely tiny, we approach the domain of quantum gravity, which is still very poorly understood. There may be a good reason for this. All extremes converge towards this point, and so it might be that only a complete theory for all physical forces, all elementary particles, and all fundamental principles together may be required to get this domain of nature properly mapped. At the same time, one may notice that then one has a description of the entire universe. This goes by the nickname "Theory of Everything". But a theory that is able to account for everything that is happening in the entire universe may well be required to possess seeds of all sources of complexity in this universe. That may be quite something to demand of a single theory. Such considerations are reasons to doubt whether ideas such as "superstring theory" contain the right mix of super-complex behavior. This speaker suspects that a lot more will be needed. I advocate procedures that are much more precise than what is presently in use. Perhaps as much as the combined intellect of all of humanity is what is needed to discover all the truths about nature in the ultimate extreme case. Individually, we are too stupid.

Complexity may well be the hardest condition in the physical world to study and comprehend. What I wished to illustrate in this talk is that most physical phenomena and laws can be pretty well understood and analyzed. Only when the circumstances become extreme, new challenges arise. I named a number of such challenges, and indicated how investigators cope with them. While they do, they will without doubt encounter new extreme conditions and new challenges that keep us all sharp. Eventually, there is reason to hope that, one day, we can combine all extremes to fit in one universal theoretical description of nature.

Origin of the Eukaryotic Cell*

Nick Lane[1]

All complex life on Earth is composed of 'eukaryotic' cells. Eukaryotes arose just once in 4 billion years, via an endosymbiosis — bacteria entered a simple host cell, evolving into mitochondria, the 'powerhouses' of complex cells. Mitochondria lost most of their genes, retaining only those needed for respiration, giving eukaryotes 'multi-bacterial' power without the costs of maintaining thousands of complete bacterial genomes. These energy savings supported a substantial expansion in nuclear genome size, and far more protein synthesis from each gene.

Keywords: Eukaryotes; Mitochondria; Bacteria; Energy per Gene; Complexity.

---

* This chapter was based on a lecture delivered by Professor Lane at the Molecular Frontiers Symposium in the Royal Swedish Academy of Sciences in May 2017.
[1] Professor of Evolutionary Biochemistry, University College London.

About Nick Lane

Nick Lane is Professor of Evolutionary Biochemistry in the Department of Genetics, Evolution and Environment at University College London.

Lane's research is on the way that energy flow has shaped evolution over 4 billion years, using a mixture of theoretical and experimental work to address the origin of life, the evolution of complex cells and downright peculiar behaviour such as sex. He was a founding member of the UCL Consortium for Mitochondrial Research, and is Co-Director of the UCL Centre for Life's Origin and Evolution (CLOE). He was awarded the 2009 UCL Provost's Venture Research Prize, the 2011 BMC Research Award for Genetics, Genomics, Bioinformatics and Evolution, the 2015 Biochemical Society Award for his outstanding contribution to molecular life sciences and the 2016 Royal Society Michael Faraday Prize and Lecture, the UK's premier award for excellence in communicating science.

Lane is the author of four acclaimed books on evolutionary biochemistry, which have been translated into 30 languages. These books treat the grand sweep of evolution from the origins of life to our own ageing and death. *Life Ascending* won the 2010 Royal Society Prize for Science Books. Lane was desorbed by *the Independent* as "One at the most exciting science writers of our time".

Lane's most recent book is entitled *The Vital Question: Why is Life the Way it Is?* (Profile/Norton, 2015). Its US subtitle is more prosaic but more self-explanatory: *Energy, Evolution and the Origins of Complex Life*. It attacks a central problem in biology — why did complex life arise only once in four billion years, and why does all complex life share so many peculiar properties, from sex and speciation to senescence? The book argues that energy has constrained the whole trajectory of evolution, from the origin of life to the properties of complex organisms including ourselves. It was named "Book of the Year" by the *Times*, *Economist*, *Wall Street Journal*, *Sunday Times*, *Independent*, *Financial Times* and *New Scientist*, and was "highly commended" by the Royal Society of Biology. Bill Gates wrote "this book blew me away".

Lane has published more than 80 peer-reviewed papers in top international journals, including *Nature*, *Science* and *Cell*, and many feature articles in magazines like

Nature, *New Scientist* and *Scientific American*. He has appeared regularly on TV and radio (including "Horizon", "In Our Time", "Radiolab", "Start the Week" and the "Today Programme"), and speaks in schools and at literary and science festivals, including New Scientist Live, the Cheltenham Festival, Hay Festival and Edinburgh Festival. He also worked for several years in the pharmaceutical industry, as Strategic Director of Medi Cine, a medical multimedia company based in London, where he was responsible for developing interactive approaches to medical education. He is a Fellow of the Linnean Society, the Royal Society of Biology and the Biochemical Society.

Thank you for the very kind introduction and thank you for the invitation. This has been an absolutely wonderful meeting, I have learned a great deal. I should say that I am not an engineer and so I feel as if I am here under slightly false pretenses. What I will talk about though is natural selection, because natural selection is a great engineer. But we heard yesterday from Frances that there are areas of biology where enzymes don't seem to have ventured. It's not that they can't do it, they can do it perfectly well. It's just that there isn't an 'incentive' for them to do it. And we see the same thing with cells and life generally. There are whole areas of possibility that life has just not explored and it's the reasons for that I find fascinating. I'm going to talk today about the origin of the eukaryotic cell. Eukaryotic cells are large complex cells with a nucleus; and eukaryote just means true nucleus, where we have most of our DNA. Essentially everything you can see, which is to say plants and animals and fungi, they're all composed of these eukaryotic cells.

So this is a three domains Tree of Life that I suspect a lot of you here will be familiar with. It goes back to Carl Woese in 1990[1]. I know a lot of people are still taught today about the five kingdoms. Actually let's just have a quick show of hands from the high school kids here. How many people have heard about the three domains and how many people..? Yeah, okay. And how many people are familiar with the five kingdoms or the six kingdoms of life? Okay, that's great. So I can tell you for nothing that education in Sweden is better than education in the UK, because I've asked this question of UK audiences and very few people have really got beyond the five or six kingdoms. So the three domains are the bacteria, the archaea and the eukaryotes. The bacteria and archaea look more or less the same, which is a key point. We've known about some archaea for hundreds of years. The methanogens for example, we've been familiar with them for 400 years. But they look the same as bacteria in their appearance. Carl Woese first started sequencing RNA and came up with this genetic Tree of Life, and the branch lengths here then give an indication of the amount of variation within these groups. This was all a bit shocking because the animals, the fungi, the plants were compressed into this small corner of the Tree of Life. This was a Copernican revolution in biology because again it pushes us into a small inconsequential corner of the universe, even of life, and it's difficult to accept but it's true. It's also difficult, at least I find it's difficult sometimes, to see things, and this is one of the reasons I enjoy writing books because that's how I come to understand the world. I try and explain, why is the Tree of Life like this? And you don't ask, I didn't ask myself these questions; but there are a couple of very strange things about this Tree of Life. Why is there so much genetic variation within these two groups? They're very different to each other, they're different in their biochemistry and all of these metabolic pathways that we've just been hearing about, they are very different, shockingly different. But in their morphology they're very similar. They've had four billion years of evolution to come up with anything in their biochemistry. It seems like they are far more sophisticated than the complex eukaryotic cells over here. They can do anything except become large and big and complex. Why didn't they do that? What was happening down here that wasn't happening in those groups? And it becomes even more strange because it's not just at the level of large plants and animals, it's at the level of single cells. So this is *Euglena*, basically it's the kind of scum that you find on any pond (Figure 1). Here is one of the more complex bacteria that you'll find. This is planctomycetes, you might just be able to see that it's got a little…looks a little bit like a nucleus. It's a kind of compartment where the DNA is, it's not very much like a nucleus, but this is one of the reasons we consider it to be quite complex; some people suggest that perhaps this is a first step towards making a nucleus. But the reason you can't see it very well is that this is roughly to scale. This is just enormously larger — on average eukaryotic cells are about 15,000 times bigger than bacteria in their volume. And you don't need to know what these are, these are actually chloroplasts in the case of *Euglena*. Here's the nucleus, you can just make out the mitochondria, but

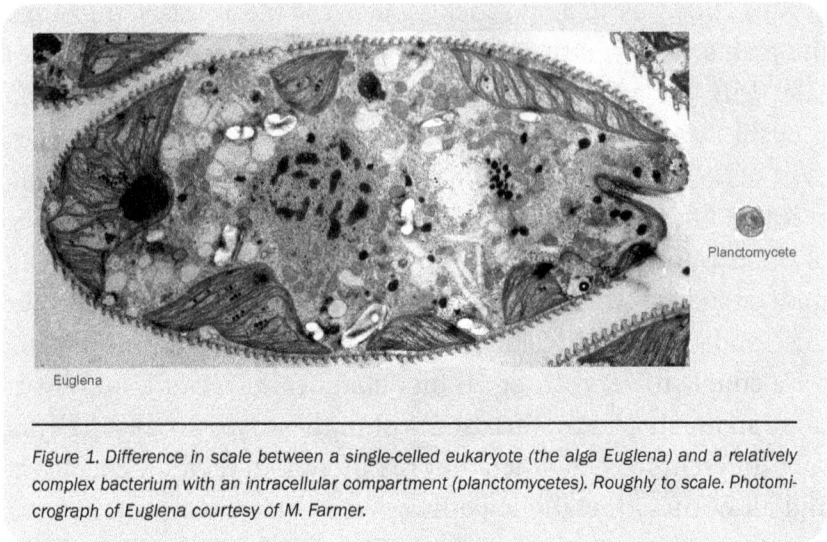

Figure 1. Difference in scale between a single-celled eukaryote (the alga Euglena) and a relatively complex bacterium with an intracellular compartment (planctomycetes). Roughly to scale. Photomicrograph of Euglena courtesy of M. Farmer.

you don't need to know what this stuff is inside a cell to appreciate that we've gone up orders of magnitude in complexity from what the bacteria and the archaea have done, and it's a puzzle.

Also at the level of eukaryotic cells, again just to make you feel even more inconsequential, this is a paramecium and this is a pancreatic acinar cell. I'm curious to know how many genes do you think a paramecium has? When I say genes, I mean genes coding for proteins. Any suggestions for how many genes paramecium has? I've had answers between 50 and a few thousand, how many do you think? Okay, well, I'll tell you the answer is 40,000. That's twice as many as we have. These are complex cells, but single-celled organisms. The reason they have so many genes is they do an awful lot of stuff inside that single cell. So, the level of complexity between different types of eukaryotic cell is very equivalent. We are really not very much more sophisticated than paramecia it seems, but we are an awful lot more sophisticated than a bacterium in terms of the morphological complexity.

So, what's going on? There's really no agreement so I'm confident in standing up here and talking to you, that whatever I say may be right or may be wrong but you'll never really know. But I think we can try to get at the problem even if we don't know the answer. These are some of the really great thinkers of biology over the 20th century. Jacques Monod we heard about yesterday, he was one of the pioneers of molecular biology. He wrote a wonderful book called Chance and Necessity in 1970. It's got quite a bleak existential view. It's an extremely exciting read but he sees the origins of life as being really very, very difficult and unlikely to happen again. One of the reasons that perhaps we don't see life elsewhere is that the origin of life is so difficult. That's what he thought. I don't think the other people in this room would necessarily think that way. But again, once evolution has got going what happens? Stephen Jay Gould wrote a great book called Wonderful Life where he imagined winding back the clock to the time of the Cambrian explosion when the first animals appeared in the fossil record, and then let it play forward again; would we end up with humans? Would we end up with even vertebrates? Or would we end up with giant octopuses on the hills or something? It's very hard to know what you might imagine. The question is what kind of engineering principles guide the evolution of life? So, on this side, these two basically think that there's a great deal of contingency, that the environment affects what happens, that the asteroid that wipes out the dinosaurs gives the mammals a chance that would never have happened if an asteroid hadn't hit in that particular place — in fact it seems perhaps vaporizing sulfates into the atmosphere by hitting specifically there in the Yucatan Peninsula. Christian de Duve, another Nobel laureate, and Simon Conway Morris believed far more in the engineering principles that underpin selection, that you will get the same things emerging time and time again because that's the best way to do it. If you want to fly you'd better have something like wings, you'd better be aerodynamic. You're going to find wings arising in bats, in birds, in insects with a rather similar structure, with rather similar aerodynamic properties for the same reasons. And so, they would argue that we would end up with something rather like the world that we have at the moment, if you were to wind back the clock to the origin of life and let it play forward again. The simple answer is we don't know who is right.

This, I think is the key problem. And I've already alluded to this. If we go back nearly 4 billion years, the kind of timeline that Gerald Joyce was talking about in the in the last talk, 3.6 to 3.8 billion years ago we see fossils in rocks that look a lot like bacteria. We don't know for sure if they are. Natural processes can give rise to shapes in rocks that look an awful lot like bacteria, but to the best of our knowledge, we see bacteria going back 3.8 billion years. A recent paper from colleagues of mine at UCL said 4 billion years[2]. I'm not sure if they're right. But, look what happened afterwards, they flat-lined for practically three and a half billion years (Figure 2). What was going on in, or what was not going on in the bacteria? Why is it that only once in this entire period do we see the origin of complex life? Now we know it happened once because all these eukaryotic cells are related to each other. We share a tremendous

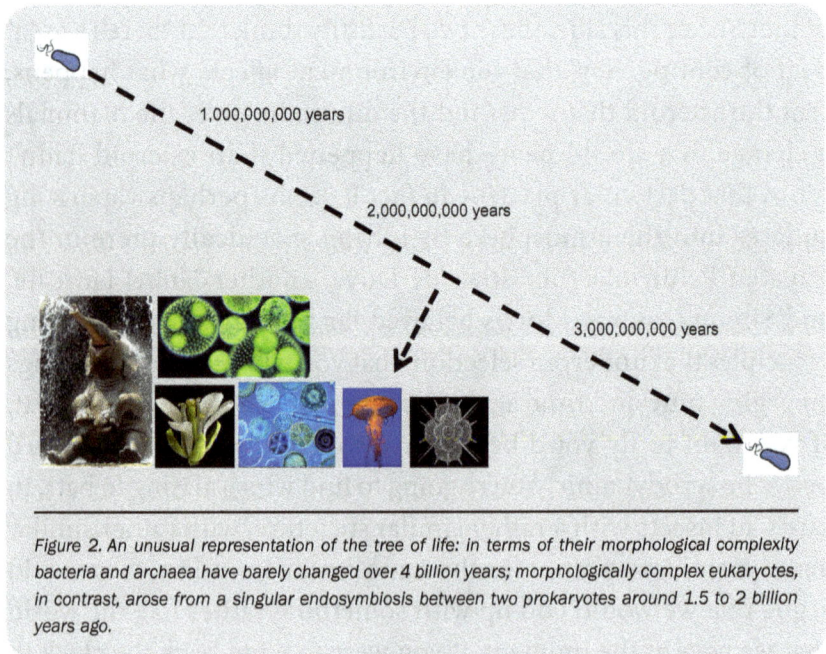

Figure 2. *An unusual representation of the tree of life: in terms of their morphological complexity bacteria and archaea have barely changed over 4 billion years; morphologically complex eukaryotes, in contrast, arose from a singular endosymbiosis between two prokaryotes around 1.5 to 2 billion years ago.*

number of traits in common. We all have a nucleus, we all have straight chromosomes, we all have genes in pieces with non-coding introns and then bits that code for proteins, we all have the same structures inside cell, we all have mitochondria and I'll say more about that. So, we know that we all share a common ancestor. And by definition that arose once; but if we look in the fossil record, we do not see things that were not eukaryotic, other origins, alternative origins of complex life. We don't see it there. We look around the world, we trawl through all kinds of muds in strange environments and we look to see alternative forms of life and we don't really find it. We find new archaea, we find new bacteria, we find amazing things, but we're not really finding different structures to cells[3].

This is one of the great evolutionary biologists and he was at UCL for a period — John Maynard Smith. He used to look for what he called the scandals of evolution. The things that really ought not to be happening like that, they should have done something different. Why did it go this way rather than that way? And this is an evolutionary scandal by his terms — all complex life is composed of eukaryotic cells. They only arose once and we all share not just the physical structure of the cells but we are all sexual. Plants are sexual and yeast are sexual too. It is right across the entire tree of eukaryotes, not just sexual, but the gametes fuse together and they go through a two-step meiosis using the same proteins, we can find the same genes right across the whole tree. It's the same. Why don't bacteria evolve any of these complex traits? They do some homologous recombination and lateral gene transfer, but they don't do two-step meiosis and they don't recombine across the entire genome[4]. It's a very different process that they do. So, the scandal is, if all of these traits evolve step by step by natural selection and each step offers some small advantage (and there's no reason to disbelieve any of that) then why is it that none of these traits arose in bacteria? It ought to be like the eye. Eyes arose essentially independently on at least 60 or 70 different occasions in different environments. A lot of these are animal eyes and so they actually do go back to a common ancestor that was a light sensitive spot on some kind of a worm, and there are some regulatory genes that they have in common, a PAX6 gene for example, but independently those regulatory genes recruited all the rest of the genes required to make an eye. And so, the octopus eye, which is here, and the human eye, they are very fine examples of convergent evolution. They're structured in essentially the same way but they evolved independently. This is *Euglena* again, here's the eye spot in *Euglena* that uses essentially the same rhodopsins that we use in our own eyes. And this is even more strange — this is a single-celled protist and its got a retina here, and its got a lens, its got a cornea. The retina is made of chloroplasts. The cornea is made of mitochondria. Its just recruited those different parts. Its got the same structure as the kind of eye that we're familiar with, but it's an utterly independent origin, this is convergent evolution. Selection would predict that we should see multiple origins of rather similar functioning things that are different to each other in different environments, different ecosystems. So why don't we see multiple origins of a nucleus, if it's a good thing to have, or sex if it's good to have sex, or phagocytosis, the ability to go around and engulf other cells (essentially eating), but we never see that in bacteria.

Well, here is a way of getting at the problem. This is a more recent tree[5]. It's actually a few years old now but I like this one. This is just the eukaryotes. And again forget about plants or algae as the main groups. There are five or six super-groups of eukaryotes. There are the excavates, the chromalveolates, the unikonts, we are all unikonts. I rather like the term although it's now becoming slightly old outmoded but we are all unikonts. Here we are, the metazoans and the fungi. The reason I like this particular tree is right at the center: this is the common ancestor of all eukaryotes, and it's rather symbolically a black hole. So there are two things to take away from this. There's far more variation within these groups than there is between the ancestors of the groups. This is what's called a Big Bang radiation. It happens apparently rather abruptly.

And that common ancestor had everything, it was a recognizable eukaryotic cell. We can trace the genomes of a large number of cells in these different groups and we can see that they share an awful lot including all these traits that I've been talking about. So we know that the eukaryotic common ancestor had all of those things, but we don't know how they arose. What we know is that bacteria don't have them. This is what I like to think of as the black hole at the heart of biology. We do not know how or why complex cells arose from bacteria. So I will put some ideas forward...these are not by any means the only ideas and we don't have facts to prove anything yet. But this looks like some kind of a bottleneck, it looks as if perhaps the conditions changed and the reason that eukaryotes suddenly took over the world is perhaps there's been a snowball earth, we know there was a snowball earth about 2.3 billion years ago and another one around 700 million years ago. This is when the entire planet froze over, we think, the geologists tell us, right down to sea level on the equator. Catastrophic global changes are undoubtedly bottlenecks that could affect tremendously the whole trajectory of life. This is another one, the Great Oxidation Event when we first start to see oxygen in the atmosphere, again from about 2.2–2.3 billion years ago, probably linked with that earlier snowball earth. These are global catastrophes and it is very easy to imagine that after this catastrophe, just as the mammals expanded after the dinosaurs, the eukaryotic cells expanded.

But that makes some predictions and people have been a little sloppy in the way that we've thought about these predictions (Figure 3). Well you might imagine if it was oxygen, for example, suddenly allowing the freedom to become bigger and more complex because now there's oxygen that animals and plants can respire and so on. You would expect multiple origins nonetheless, you would expect that the cyanobacteria, the photosynthetic bacteria, would give rise to photosynthetic plants and algae. You would expect that osmotrophic bacteria would give rise to fungi, putting enzymes out into the surrounding area, breaking down food and then taking up the monomers. You would expect separate origins from the

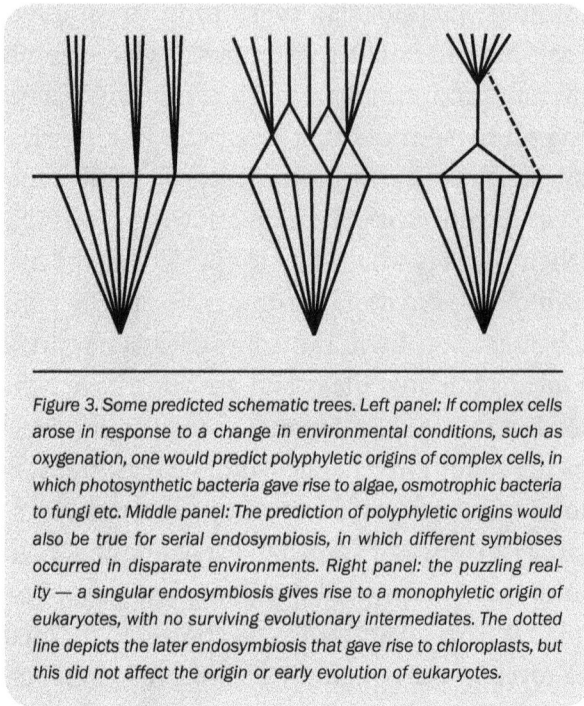

Figure 3. Some predicted schematic trees. Left panel: If complex cells arose in response to a change in environmental conditions, such as oxygenation, one would predict polyphyletic origins of complex cells, in which photosynthetic bacteria gave rise to algae, osmotrophic bacteria to fungi etc. Middle panel: The prediction of polyphyletic origins would also be true for serial endosymbiosis, in which different symbioses occurred in disparate environments. Right panel: the puzzling reality — a singular endosymbiosis gives rise to a monophyletic origin of eukaryotes, with no surviving evolutionary intermediates. The dotted line depicts the later endosymbiosis that gave rise to chloroplasts, but this did not affect the origin or early evolution of eukaryotes.

cells that are best pre-adapted to the new conditions, but that's not what we see. The Serial Endosymbiosis Theory from Lynn Margulis again, she anticipated that there would be multiple different types of endosymbiosis[6], different types of bacteria interacting with each other, and so multiple different origins of complex life, but that's not what we see[7]. What we seem to see is something more like this, and I shall say more about this. It does not look like an environmental bottleneck but a structural bottleneck, something to do with the structure of cells, where one cell got inside another one (Figure 3). This is the black hole area where everything evolved, and this is the moment where everything takes off into the modern groups that we see. This is the acquisition of the chloroplasts here — the cyanobacteria that gave rise to the algae. They didn't change the fundamental direction of evolution. But here is what I shall talk about for the remainder of the talk.

The one thing which has really changed over the last 20 years, and again this is grounded in medical research because most of these cells, once called archezoa, are in fact parasites of one sort or another. It was thought that none of these cells have mitochondria, and they all looked rather morphologically primitive. The assumption was that the archezoa were early branching eukaryotes that would give us an indication of how complex life arose. Well it turns out after a lot of studying by a number of groups over many years that they all do have mitochondria, just not as we know them. They've become what are called relict organelles, although they were lost altogether in the case of *Monocercomonoides*[8], but they all had them once and they became specialized for different tasks. So we now know that the common ancestor of eukaryotes already had mitochondria. And when we start looking at the genomes then we can see that potentially the origin of eukaryotes and the acquisition of mitochondria were one and the same thing. The mitochondria, in case you don't know, they are the powerhouses of eukaryotic cells. These are our own mitochondria here. And what's going on here was discovered back in 1961 by Peter Mitchell, who called it chemiosmotic coupling[9]. This is Peter Mitchell in 1947 with Jennifer Moyle, his long term collaborator, through all their lives really (Figure 4). Mitchell won the Nobel Prize in 1978 for his visionary ideas. Jennifer Moyle had been the experimentalist who tested most of these ideas and showed that they were essentially true, that this really is how cells work. It's interesting to me, perhaps in this arena especially, to wonder about how one balances between the ideas and the experiments. Mitchell received the Nobel Prize because he had developed the ideas himself, but he didn't really do the experiments. If Jennifer Moyle had not done those experiments nothing would have proved Mitchell to be correct, and so perhaps they deserved it together, I don't know. But this is what they showed, this is what's happening in you right now. Electrons are being stripped from food and they're being passed down the respiratory chain (Figure 4). This is the membrane here, the cristae membranes in here. These are giant protein complexes of the respiratory chain and I'm just symbolizing them as small balloons. What we have is a current

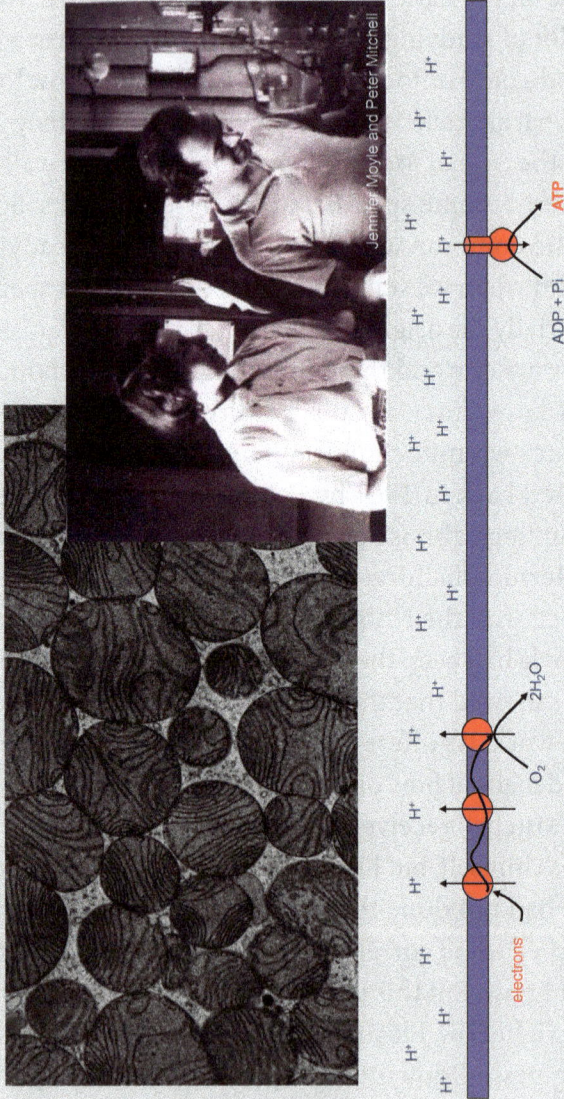

Figure 4. Mitochondria showing the cristae membranes where respiration takes place. The mechanism of respiration, known as chemi-osmotic coupling, was elucidated by Peter Mitchell in collaboration with Jennifer Moyle and others. The cartoon shows the mechanism of chemiosmotic coupling, in which the flow of electrons from 'food' to oxygen within the membrane powers the extrusion of protons across the membrane; the flow of protons through the ATP synthase, a kind of turbine, powers the synthesis of ATP, the universal energy currency of life.

of electrons from food to oxygen, and that current of electrons is powering the extrusion of protons across the membrane. So we end up with a kind of reservoir of protons on this side of the membrane and relatively few on that side. Protons have a positive charge which means you now have an electrical potential across this membrane. But there's also a concentration difference. This is what Peter Mitchell called the proton-motive force. And this, the ATP synthase, is a wonderful machine, it's a rotating motor, essentially equivalent to a turbine in a hydroelectric dam. This is equivalent to water flowing through the turbine: the protons flowing through the ATP synthase are turning the motor, which is powering the synthesis of ATP, and that's powering everything else in the cell. So this is what's happening in respiration. Now the mitochondria were bacteria once, we're pretty certain about that, in fact there's not really any serious opposition to that idea anymore. That goes back to Lynn Margulis, in fact it goes back much before Lynn Margulis, but she was the person who nailed the idea in 1967, 50 years ago this year[6]; there's a celebratory issue of The Journal of Theoretical Biology where she published that original paper coming out shortly. Incidentally, she was married to Carl Sagan — they must have had some pretty amazing breakfast time conversations, I should imagine, in that household. They unfortunately divorced before 1967 when she first published that paper. But the key point is that mitochondria were bacteria once, and that's firmly established.

This is an alternative Tree of Life going back to 1992, about the same time as Carl Woese and this is a different way of seeing it[10]. This is Jim Lake. So this is the Carl Woese' Tree of Life, the three domains Tree of Life. And Jim Lake said no, no, it's not, it's a ring; apparently he sent a paper to Nature entitled One Ring to Rule Them All. And they rejected the title unfortunately but published the paper, and this was the essence of it. He was looking at where these genes came from. And a large number come from bacteria, a large number come from archaea, and it's this genomic fusion that is giving rise to the eukaryotes. This was a radical idea that nobody really believed for quite a long time but over the last maybe six or seven years it's become clear from phylogenetics that something like this is indeed the case. So this is the classic three domains tree with the eukaryotes at the top. These are different groups of archaea here and the bacteria down at the bottom. That's the classic tree. What we see now is that the eukaryotes branch inside the archaea[11]. This is from concatenated sequences of 40 to 60 genes, the more genes you have, the stronger the signal, the trouble is the more genes you have the more likely to be passed around by lateral gene transfer they are, and that produces noise which confounds the signal. So it's a difficult balance but this has been repeatedly found in a lot of studies now. All this means the host cell was an archaeon. We don't know what kind of archaeon but we're getting closer. This is a paper from earlier this year and some earlier work from a couple of years ago. This is the Asgard superphylum. The Lokiarchaeota were discovered at Loki's Castle a couple of years ago[12]. We don't know what they look like. This is just metagenomic screening of the muds around there. And these are the most similar genomes to eukaryotes — here are the eukaryotes branching. There's now several groups so we have the Lokiarchaeota and then the Odinarchaeota, the Thorarchaeota, the Heimdahlarchaeota, we've got all these Nordic gods. We don't know what any of these cells look like, we only have the genome sequences and they're all relatively small, they're all 4,000–5,000 genes or so. So they're kind of standard size for archaeal genomes. And they have some interesting properties. They seem to have a pretty dynamic cytoskeleton compared with other archaea. They seem to be capable of some membrane remodeling, but we haven't seen it. We don't really know if are they slightly phagocytic, can they begin to engulf other cells, or is it just how they divide in half where you also need to change the membrane structures. So, we don't know yet what kind of archaea, but we are fairly sure that we have something here, this, something within the Asgard phylum, the Lokiarchaeota acquires a bacterium.

And this is where it becomes really quite difficult and I'd just like to show you. This is a paper from last year from Nature which gives an indication of how difficult it is to get at some of these problems[13]. This is looking at the stem length of genes that come from bacteria. There's a lot of genes that come from bacteria, some of them branch with the alphaproteobacteria. These are definitely the mitochondrial genes. Here are the genes that are definitely with the mitochondria. But these genes branch with the bacteria but not clearly with any particular group. And the stem length is longer which means to say there are more differences. This structure is the nucleus, these are bacterial genes that you find in the nucleus, and they are more distantly related. These are the endomembrane systems, these are the mitochondria. So what they said is that the number of differences represent time roughly, basically a molecular clock. There's an assumption that the number of differences accumulating over time gives an indication of the time that has passed. If that's the case then the nucleolus, the nucleus, the endomembrane systems all arose before the acquisition of mitochondria. And if this is correct then everything I'm going to tell you in the rest of the talk is incorrect. That's the kind of thing that troubles you when you're trying to sleep at night. Are they right? Have I been wasting my time for the last 10 years? Well, they may be correct, I don't know. They may also be wrong. The question is, do I have an explanation as to why they would see that. I assume their data is correct (though I know it has been challenged). So, do I have an answer? Well, yes, I do; I don't know if my answer is correct but it helps me to sleep. Let me explain.

This is an old tree going back now to 1998[14]. Here are the bacteria, the archaea and these groups, these are the cells that I showed you before, the Archezoa. These are cells that don't have mitochondria. We used to think that they were early branching. Why? Well, they're simple in their morphology but also they have these long, long branch lengths. And that means that the place they branch on the tree is an artifact called long branch attraction. They're shown to branch here and here was the acquisition of mitochondria, actually right up there somewhere. So the length of these branches does not give an indication of the amount of time, it gives you an indication of the amount of evolution that has happened, the number of changes. And we can't constrain the time necessarily with that. So, this tree is an artifact and everybody's agreed about that now. And I think that is probably the best way to think about this other study.

This is a painting by Odra Noel, a beautiful painting of a eukaryotic cell. Here is the nucleus, here are the mitochondria, here are these endomembrane systems that I was talking about. Why would you have more evolution in the endomembranes or the nucleus? Well, the genes in the mitochondria are doing what they always did. They're doing respiration in a mitochondrial setting. They are under strong purifying selection for the same job in the same setting that they always did, it never changed. And purifying selection means that you have fewer changes in sequence because changes get eliminated if they're not helpful to you. But the endomembranes, well, they don't exist in bacteria really. And the nucleus, it doesn't exist in bacteria. So there must have been, theoretically there must have been a period of strong adaptive selection and adaptive selection by its nature is forcing changes on you. You're changing to a new purpose, a new function and so you're going to have lots of changes. And that's going to increase the branch length.

If I'm right, what does it say? Well, this is the origin of the eukaryotic cell. These are bacteria living inside a bacterial cell (Figure 5). This is the only example that we know of, of bacteria inside a free-living bacterial cell[15]. There's plenty of bacteria inside eukaryotic cells which are large and complex and often engulf cells like bacteria for a living; but bacteria don't do that. This is a cyanobacterium, these are thylakoid membranes, it has a cell wall. It did not engulf those cells by phagocytosis, but we don't know how they got there. But what good is it if those cells inside went on to become the mitochondria? Why was the acquisition of mitochondria any use? It seems reasonable that we should look for

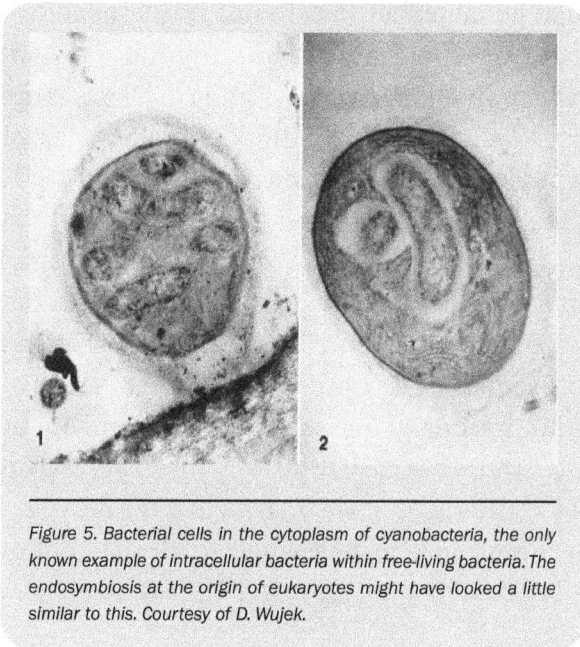

Figure 5. Bacterial cells in the cytoplasm of cyanobacteria, the only known example of intracellular bacteria within free-living bacteria. The endosymbiosis at the origin of eukaryotes might have looked a little similar to this. Courtesy of D. Wujek.

the answer in terms of energy in one way or another. Well, they are the power packs, they produce the ATP so that's where we should look for an answer. But, if you look, these are values taken from the literature[16]. This is the metabolic rate of bacteria compared with single-celled protists. This is a log scale in each case so the bacteria respire about three times faster than single celled eukaryotes. So, it's not the case that they help us to respire faster, it's not as simple as that; but that's per gram. If we look per cell, it changes around. So, this is a log scale again. Now here we can see that a single eukaryotic cell consumes about 5,000 times more oxygen per minute than a single bacterial cell[16]. Why? Because they're 15,000 times larger — of course they do, they're juggernauts. This is a silly comparison in one sense, but it's beginning to get at the problem. What are they spending all this energy all on? Well, this is energy per mega base of DNA. And you can see it's roughly similar again in this case. So, eukaryotic cells are becoming a lot larger, they have a lot more energy to spend and they're spending it on maintaining a much larger genome; but they're not really spending any more per mega-base of DNA than in a bacterium does. What are they actually spending it on? Well, this is some old work. This is not the kind of thing that many people think about anymore, but Frank Harold did some lovely work in the 1970s on the ATP budget of bacteria[17]. The answer is to a large degree protein synthesis. That's not true of us, that's not true of multicellular organisms generally. But it seems to be true of many protists and bacteria that 75 to 80 percent of the energy budget of a cell goes on protein synthesis. Here is DNA synthesis, it's a trivial cost in comparison. There can be lots of futile cycling going on in an environment but this is in growing cells. And when they're growing the biggest costs are to do with protein synthesis. So if we then look at the energy availability per protein coding gene and we equalize for the number of genes then we find again a roughly 5,000 fold difference (Figure 6). That's

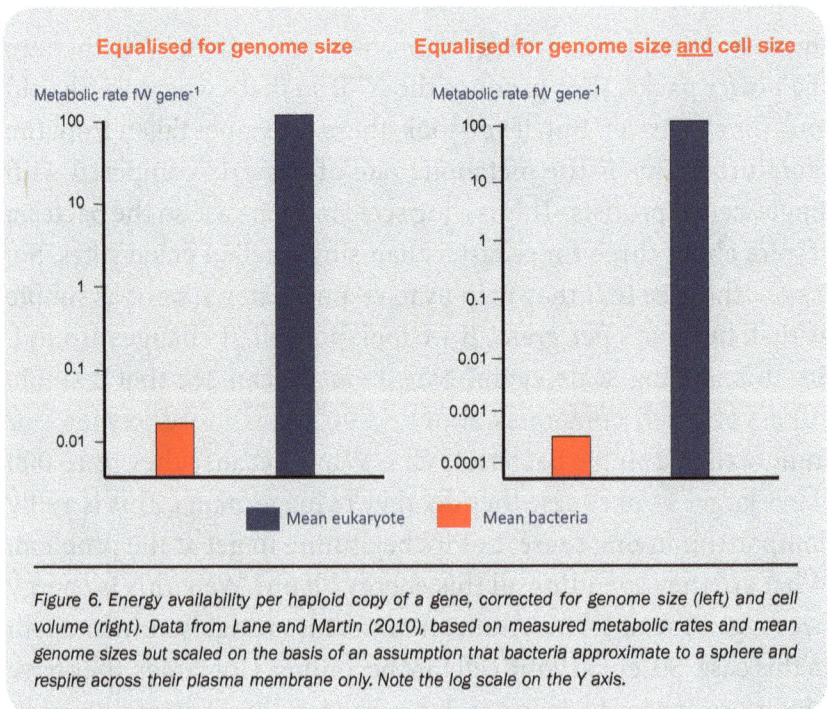

Figure 6. Energy availability per haploid copy of a gene, corrected for genome size (left) and cell volume (right). Data from Lane and Martin (2010), based on measured metabolic rates and mean genome sizes but scaled on the basis of an assumption that bacteria approximate to a sphere and respire across their plasma membrane only. Note the log scale on the Y axis.

to say, a single eukaryotic cell has about 5,000 times more energy per gene than a single bacterial cell. That does not mean that they should have 5,000 times more genes, it means that they can do 5,000 times as much gene expression. They can make 5,000 times as many proteins — because they're 15,000 times larger of course they have to. It's a statement of the obvious really. So we need to correct for cell volume as well. But now if we correct for cell volume then we get this massive difference opening up — this equates to a 200,000 fold difference in energy per gene when we've corrected for the gene number and cell volume (Figure 6)[16]. Now really this is silly because we've made a silly assumption underpinning it, but it's interesting to get it why it's a silly assumption. So bacteria pump protons in exactly the same way that our mitochondria do. They're pumping them across the plasma membrane. And if you increase the volume of the cell, then of course you have surface area to volume constraints. ATP synthesis depends on the surface area, and protein synthesis depends on the volume. That's where that number of 200,000 came from, and it's a silly number because we know that bacteria can internalize membranes and get around that problem immediately (Figure 7). So why don't they actually do that; or do they do that? Well, to an extent they do. Let's see what the problem is.

Here is another paper that seems to be at odds with my position. This is from Mike Lynch[18]. And what he's plotted here is the number of ATPase enzymes against the surface area. You can see a nice straight line, bacteria down here, eukaryotes up there. So it basically correlates beautifully with surface area — exactly what you would expect. You have more surface area, you have more energy, more ATPase enzymes. And this is the number of ribosomes in a cell against the cell volume. Again, the bigger the cell, the more ribosomes. So you would say that these are just continuous and really there's nothing special about eukaryotes. But what's concealed here in these log scales is that all the bacteria are down here, in every case; we've got one eukaryote there but by and

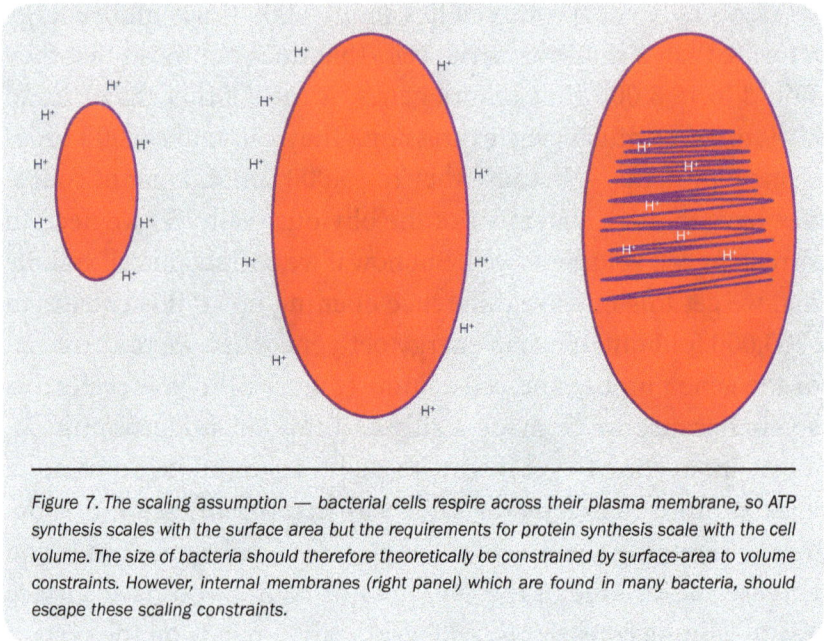

Figure 7. The scaling assumption — bacterial cells respire across their plasma membrane, so ATP synthesis scales with the surface area but the requirements for protein synthesis scale with the cell volume. The size of bacteria should therefore theoretically be constrained by surface-area to volume constraints. However, internal membranes (right panel) which are found in many bacteria, should escape these scaling constraints.

large all the eukaryotes are up here, all the bacteria down there, and there's a couple of orders of magnitude difference between the largest bacteria and the smallest eukaryotes, there's practically no overlap. I would say these are two different continua, and there is something else stopping bacteria from expanding their surface area up to eukaryotic proportions. Look here (Figure 8). These are bacteria: cyanobacteria, nitrogen fixing bacteria; and these are the internal membranes — they can internalise membranes perfectly easily. And this is paramecium again — these are the mitochondria in paramecium. The difference is scale again. These are roughly to scale. And this is only a small section of a paramecium. It's expanded up over orders of magnitude. The difference here is that these mitochondria all have genes of their own. They started out as bacteria, they lost most of their genes, but they ended up retaining always a very similar subset of genes. They seem to need those genes to control respiration. And so here we're going to converge on engineering principles in one sense. You need genes to control

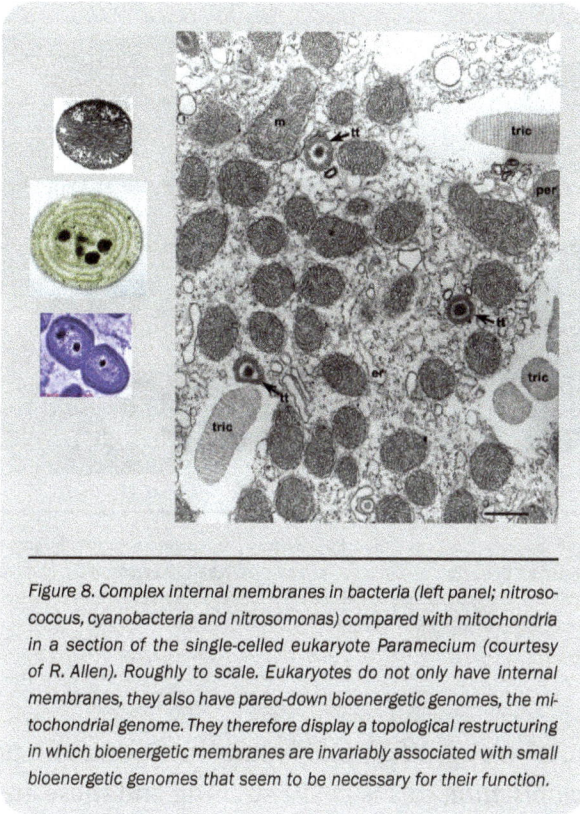

Figure 8. Complex internal membranes in bacteria (left panel; nitroso-coccus, cyanobacteria and nitrosomonas) compared with mitochondria in a section of the single-celled eukaryote Paramecium (courtesy of R. Allen). Roughly to scale. Eukaryotes do not only have internal membranes, they also have pared-down bioenergetic genomes, the mi-tochondrial genome. They therefore display a topological restructuring in which bioenergetic membranes are invariably associated with small bioenergetic genomes that seem to be necessary for their function.

respiration[19]. If you want to get bigger then you just have more mitochondria. Each one comes with its own regulatory system in these genes. So if they want to expand up to eukaryotic size, well perhaps bacteria can't because they don't have the genes. That's easily tested because there are some giant bacteria around[20]. This is *E. coli* here, here's paramecium. Paramecium is dwarfed by this battleship of a cell. This is *Epulopiscium*, it's a bacterial cell and you can see it with the naked eye (Figure 9). This is even larger — *Thiomargarita*. It's basically a giant vacuole with a thin film of cytoplasm surrounding it; this is a single bacterial cell and this is *Drosophila* the fruit fly. So *Thiomargarita* is almost as big as the head of *Drosophila*, it really is a monstrous cell. If you need genes to control respiration and this cell is respiring across this plasma membrane then there better be a lot of genes there, otherwise these

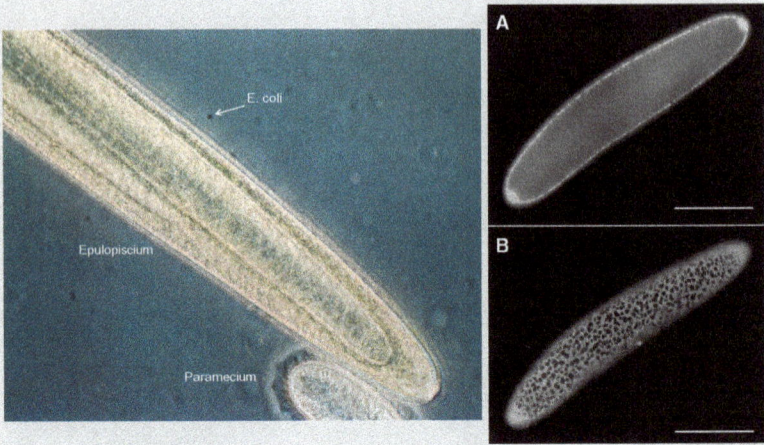

Figure 9. An example of a giant bacterium, Epulopiscium, with extreme polyploidy (shown with DAPI staining, right panel). Epulopiscium can have as many as 200,000 copies of the complete genome, positioned right next to the bioenergetic plasma membrane, as predicted. Extreme polyploidy has a huge energetic cost. Courtesy of E. Angert.

ideas are just wrong. Well, this is known as extreme polyploidy. This is *Epulopiscium*, this is DAPI staining and these are copies of the complete genome (Figure 9).

When I first saw this picture I realized that maybe there's something in what we're talking about. There are 200,000 copies of the complete genome[20]. Each genome is three mega bases of DNA, so 3 million base pairs. And this is Thiomargarita, this is the giant vacuole, this is the thin film of cytoplasm, and here there's about 15–20,000 copies of the complete genome. We can add up all of that genomic weight and ask what's the energy available per gene per haploid copy of each gene. This is *E. coli*, *Thiomargarita*, *Epulopiscium*: the energy per gene is essentially the same. What they are is a kind of a consortium of bacteria that are fused together, and each genome is controlling a similar volume of cytoplasm and a similar area of plasma membrane. No doubt there are other advantages to being larger but they have got nothing to do with energy. So why then are eukaryotes different,

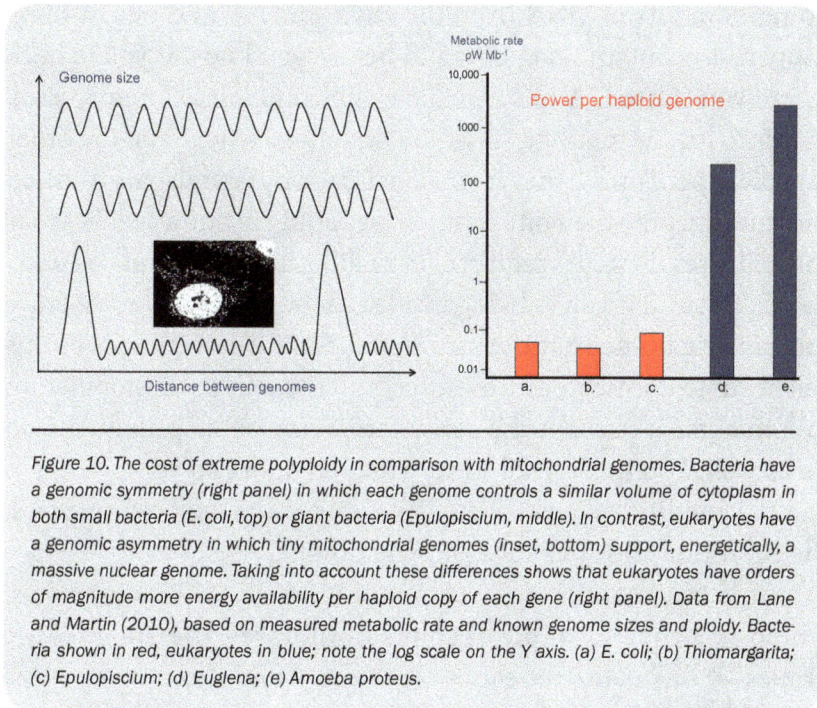

Figure 10. *The cost of extreme polyploidy in comparison with mitochondrial genomes. Bacteria have a genomic symmetry (right panel) in which each genome controls a similar volume of cytoplasm in both small bacteria (E. coli, top) or giant bacteria (Epulopiscium, middle). In contrast, eukaryotes have a genomic asymmetry in which tiny mitochondrial genomes (inset, bottom) support, energetically, a massive nuclear genome. Taking into account these differences shows that eukaryotes have orders of magnitude more energy availability per haploid copy of each gene (right panel). Data from Lane and Martin (2010), based on measured metabolic rate and known genome sizes and ploidy. Bacteria shown in red, eukaryotes in blue; note the log scale on the Y axis. (a) E. coli; (b) Thiomargarita; (c) Epulopiscium; (d) Euglena; (e) Amoeba proteus.*

why is a cell within a cell an advantage? Well, these cells went on to become the mitochondria and the difference is they are not genomes (Figure 10). They are bacteria, they are autonomous self-replicating cells in populations capable of undergoing selection and change over time, and that's exactly what happened[16]. What happens in bacteria generally? Well, imagine the yellow cell here is a cell that lost a gene that it doesn't need. Let's say it's for respiring with lactose or something like that. It doesn't need that gene and if it loses it, it cuts out a bit of DNA. It will grow a little bit faster and there's plenty of evidence showing that gene loss is an important factor in bacterial evolution. Over time most of the cells in that population will lose that bit of DNA that they don't really need because now they can grow slightly faster. But then the conditions change. Suddenly the environment is swamped with lactose again so the cells need that gene again. What do they do? Well, they pick up random bits of DNA from the environment, and one of these happens to contain that gene, and before you know it you're back where you started. And so we have opposing selection processes. One is to lose as many genes as you can afford to lose and the other is to pick them up whenever you need them by lateral gene transfer. But now imagine the same thing is happening inside a cell. Let's say this cell loses the genes required for making a cell wall and you don't need a cell wall if you're living inside another cell, you're in a fairly unchanging homeostatic environment. So the descendants of this cell begin to dominate and eventually all the cells in the population lose that gene, perhaps the genes for a bacterial flagellum and so on as well. You don't need those genes anymore so you lose them and the conditions never change. As long as the host cell survives, the conditions don't change and you do fine.

This trajectory is really common in bacteria. This is typhus, which is a nice example. Typhus obliterated the Napoleonic armies — this is the retreat of Napoleon from Moscow. And this is *Rickettsia*, the cause of Typhus. It is transmitted by the flea and has lost most of its genes — it's now down around about one mega base of DNA, so about a quarter of the size of *E. coli*. That trajectory is really common. This is the range of genome sizes of free living bacteria, which goes up to about 12 or 13 mega bases of DNA. Again, there is a continuum with eukaryotes, but eukaryotes go up to 150,000 mega bases, so orders of magnitude more[21]. And the obligate symbionts and endosymbionts are down here, one mega base or less. We know hundreds of examples of bacterial cells living inside eukaryotes and virtually all of them have undergone this genomic streamlining. Why is that useful? Well, it's useful because the bacteria still make as much ATP as they ever did, it's just that their overhead costs are being constantly lowered. Imagine that you've got a hundred endosymbionts. This is a silly thought experiment but it gives an indication of the size of the advantage[22]. So imagine you've got a hundred endosymbionts, a hundred bacteria

Energy savings	ATP cost	Actin costs	ATP cost
5% of 4 Mb genome = 200 proteins	200	Length of monomer	29 nm
100 endosymbiont genomes	100	Monomers per micron	35
2000 copies of each protein	2000	374 residues per monomer	374
250 amino acids per protein	250	Dimers in actin filament	2
5 ATPs per peptide bond	5	5 ATPs per peptide bond	5
Total per 24 hr lifecycle	50 x 10⁹	Total per micron of actin	131,000
Total per second	580,000	Microns per second	4

Figure 11. Table showing energy savings as a result of 5% gene loss from 100 endosymbionts (left column) and scale of advantage in terms of de novo synthesis of actin filaments, the basis of the dynamic eukaryotic cytoskeleton. The energy savings from not needing to synthesise proteins from 200 genes in 100 endosymbionts equates to 580,000 ATPs per second, assuming a 24-hour lifecycle; in principle this is enough to power the de novo synthesis of 4 micometers of actin per second.

living inside this cell, and each of them has a standard bacterial sized genome with four mega bases of DNA, so about 4,000 genes (Figure 11). And let's assume that they lose 200 of these genes, which they don't need. What are the energy savings of not making those proteins? Well, it is losing 200 genes from 100 endosymbionts, and in bacteria each gene would normally produce about 2,000 copies of the protein that it encodes. On average in bacteria, there's about 250 amino acids in a single protein and the ATP costs are round about five ATPs per peptide bond. So we have a total cost of 50 billion ATP to make those proteins' or the equivalent cost savings if we don't make those proteins. If you translate that into a 24 hour life cycle that would be 580,000 ATPs per second of energy savings. What could you spend it on? Well, imagine a dynamic cytoskeleton, an actin cytoskeleton, which is one of the things that sets eukaryotes apart. What are the costs of that? Actin is made up of a series of globular proteins which are joined together into a filament. And there's two filaments wrapped around each other. The length of the monomer, the single globular protein subunit, is 29 nanometers, which means there are 35 of them in a micrometer. There's 374 amino acids in each of these monomers and again we assume 5 ATPs per peptide bond. So it would cost 131,000 ATPs to make one micrometer of actin, which means you could make 4 microns per second for those energy savings. And so you see that gene loss from the endosymbionts produces so much superfluous energy that eukaryotes were effectively just swamped in ATP. That makes all the difference in the world in terms of what they can do. And this shows that difference. We've already seen this, but this is now measured numbers (Figure 10). This is not theoretical scaling of a sphere. This is the known metabolic rate, the measured metabolic rate, the known genome size, the known polyploidy. And this is a log scale, there are 3 or 4 orders of magnitude difference between bacteria and *Euglena* or large amoeba such as *Amoeba proteus*. We have far more energy availability per gene than bacteria do. Again, there's an overlap between bacteria, archaea and eukaryotes — this is genome size down here, a log scale again, where as up here near the top we see mammals — pretty much anything goes; we can support a genome as large as we want through this method but bacteria and archaea never really get above here. So 3 or 4 orders of magnitude difference again, it comes out exactly the same in that sense.

To finish then, this is the defining signature of eukaryotes and something that we have to wrestle with when we're thinking about treating diseases. Bacteria have a kind of genomic symmetry (Figure 10). Each cell has a similarly sized genome controlling a similar volume of cytoplasm and a similar area of cell membrane[23]. If you were to take a random walk through a population of *E. coli* you would find each cell has a similar size genome. You'd find the same thing if you walk through the cytoplasm of *Thiomargarita* — you'd keep finding similarly sized genomes controlling similar volumes of cytoplasm. But if you were to do that through a population of *Euglena* you'd find a massive nuclear genome supported energetically by these tiny mitochondrial genomes. And so we have a genomic asymmetry. We don't have really a single human genome. There are two human genomes — the nuclear and the mitochondrial genomes. The interactions between the two genomes are tremendously important to human health.

So this is the final slide: why did complex life only arise once? Well it's very difficult to get one cell inside another cell. When we're talking about bacteria, we know of one example, possibly a couple of others that are more equivocal. There must have been thousands of examples over evolutionary time. So it's a bottleneck but it's not a very tight bottleneck. But once these bacteria got inside, we have effectively a simple bacterial cell with other bacterial cells living inside it (Figure 12). This is why we don't see any intermediates — I would say we need to be looking at the interactions between the host cell and the endosymbionts to explain a great deal of eukaryotic complexity[3]. This is the tightest part of the bottleneck — very few cells survive the conflicting demands of living in an intimate union, through adaptations such as sex. We can examine these selective forces by standard population genetics, and we can model some of the outcomes, and make predictions and test those predictions; but I'm not going to talk about that now. I'm just going to stop now and say thank you very much.

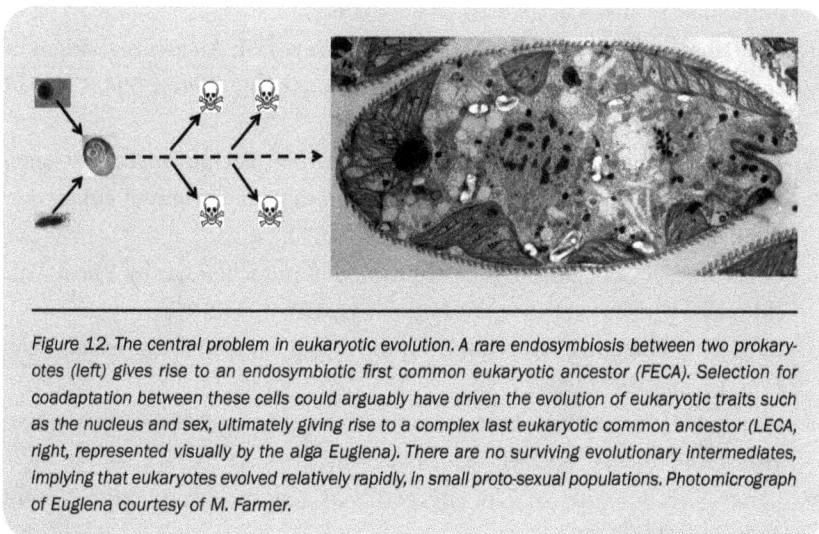

Figure 12. The central problem in eukaryotic evolution. A rare endosymbiosis between two prokaryotes (left) gives rise to an endosymbiotic first common eukaryotic ancestor (FECA). Selection for coadaptation between these cells could arguably have driven the evolution of eukaryotic traits such as the nucleus and sex, ultimately giving rise to a complex last eukaryotic common ancestor (LECA, right, represented visually by the alga Euglena). There are no surviving evolutionary intermediates, implying that eukaryotes evolved relatively rapidly, in small proto-sexual populations. Photomicrograph of Euglena courtesy of M. Farmer.

References

1. Woese, C., Kandler, O. & Wheelis, M. Towards a natural system of organisms: Proposal for the domains archaea, bacteria, and eucarya. *Proc. Natl. Acad. Sci. U.S.A.* **87**, 4576–4579 (1990).

2. Dodd, S., Papineau, D., Grenne, T., Slack, J., Rittner, M., Pirain, F., O'Neil, J. & Little, C. Evidence for early life in Earth's oldest hydrothermal vent precipitates. *Nature* **543**, 60–64 (2017).

3. Lane, N. *The Vital Question: Why is Life the Way it Is?* (Profile Books, London, 2015).

4. Martin, W. Mosaic bacterial chromosomes: A challenge en route to a tree of genomes. *BioEssays* **21**, 99–104 (1999).

5. Koonin, E. The origin and early evolution of eukaryotes in the light phylogenomics. *Genome Biol.* **11**, 209 (2010).

6. Sagan, L. On the origin of mitosing cells. *J. Theor. Biol.* **14**, 225–274 (1967).

7. Lane, N. Serial endosymbiosis or singular event at the origin of eukaryotes. *J. Theor. Biol.* **434**, 58–67 (2017).

8. Karnkowska, A., Vacek, V., Zubacova, Z., Treitli, S., Petrzelvcova, R., Eme, L. *et al.* A eukaryote without a mitochondrial genome. *Curr. Biol.* **26**, 1274–1284 (2016).

9. Mitchell, P. Coupling of phosphorylation to electron and hydrogen transfer by a chemiosmotic type of mechanism. *Nature* **191**, 144–148 (1961).

10. Rivera, M. & Lake, J. Evidence that eukaryotes and eocyte prokaryotes are immediate relatives. *Science* **257**, 74–76 (1992).

11. Williams, T.A., Foster, P.G., Cox, C.J. & Embley, T.M. An archaeal origin of eukaryotes supports only two primary domains of life. *Nature* **504**, 231–236 (2013).

12. Zaremba-Niedzwiedzka, K., Caceres, E.F., Saw, J.H., Bäckström, D., Juzokaite, L., Vancaester, E. *et al.* *Asgard archaea* illuminate the origin of eukaryotic cellular complexity. *Nature* **541**, 353–358 (2017).

13. Pittis, A.A. & Gabaldon, T. Late acquisition of mitochondria by a host with chimaeric prokaryotic ancestry. *Nature* **531**, 101–104 (2016).

14. Katz, L.A. Changing perspectives on the origin of eukaryotes. *Trends Ecol. Evol.* **13**, 493–497 (1998).

15. Wujek, D.E. Intracellular bacteria in the blue-green-alga *Pleurocapsa minor*. *Trans. Am. Microsc. Soc.* **98**, 143–145 (1979).

16. Lane, N. & Martin, W. The energetics of genome complexity. *Nature* **467**, 929–934 (2010).

17. Harold, F.M. *The Vital Force: A Study of Bioenergetics* (Freeman, 1986).

18. Lynch, M. & Marinov, G.K. Membranes, energetics, and evolution across the prokaryote-eukaryote divide. *eLife* **6**, e20437 (2017).

19. Allen, J.F. The CoRR hypothesis for genes in organelles. *J. Theor. Biol.* **434**, 50–57 (2017).

20. Angert, E.R. DNA replication and genomic architecture of very large bacteria. *Annu. Rev. Microbiol.* **66**, 197–212 (2012).

21. Elliott, T.A. & Gregory, T.R. What's in a genome? The C-value enigma and the evolution of eukaryotic genome content. *Philos. Trans. R. Soc. B* **370**, 20140331 (2015).

22. Lane, N. Bioenergetic constraints on the evolution of complex life. *Cold Spring Harb. Perspect. Biol.* **6**, a015982 (2014).

23. Lane, N. Energetics and genetics across the prokaryote-eukaryote divide. *Biol. Direct* **6**, 35 (2011).

Synthetic Tissues*

Robert Langer[1]

By synthesizing new polymeric materials and combining them with growth factors or cells, new tissues and organs can potentially be created for use in drug testing-thereby potentially reducing animal and human testing- and to treat disease. Examples discussed include blood vessels, heart muscle, spinal cord repair, artificial skin, cartilage, and pancreas.

Keywords: Tissue Engineering; Polymer; Vascular Biology.

---

* This chapter was based on a lecture delivered by Professor Langer at the Molecular Frontiers Symposium in the Royal Swedish Academy of Sciences in May 2017.
[1] Department of Chemical Engineering, Massachusetts Institute of Technology, Cambridge, MA 02139-4307, USA.

About Robert Langer

"Our work is at the interface of biotechnology and materials science. A major focus is the study and development of polymers to deliver drugs, particularly genetically engineered proteins and DNA, continuously at controlled rates for prolonged periods of time. Our interest in drug delivery systems has extended to selective drug or substance removal systems that may circumvent toxicity. In addition, we are developing drugs that specifically inhibit the process of neovascularization that is critical to several disease processes without interfering with existing blood vessels. Finally, we have been involved in creating approaches to engineer new tissues. In particular, we are synthesizing new biodegradable polymer systems to be used in mammalian cell transplants to create liver, cartilage, and nerves, and are developing bioreactors for these purposes."

Robert S. Langer is the David H. Koch Institute Professor at MIT and, as one of 13 Institute Professors, has received the highest honor that can be awarded to a faculty member. His research is at the interface of medicine — cancer in particular — and materials science and chemical engineering. Nanotechnology is a major focus area for Langer, who is developing new nanoparticles to treat cancer and other diseases. Specifically, he designs polymer, lipid, and polymer-lipid hybrid nanocarriers for improved drug delivery, as well as similar controlled delivery systems for genetically engineered therapeutic proteins, DNA and RNA. Langer's work also includes the creation of novel approaches for the engineering of new tissues and organs.

Langer has written over 1,400 articles. He also has nearly 1,300 patents worldwide. Langer's patents have been licensed or sublicensed to over 350 pharmaceutical, chemical, biotechnology and medical device companies. He is the most cited engineer in history. He served as a member of the United States Food and Drug Administration's SCIENCE Board, the FDA's highest advisory board, from 1995–2002 and as its Chairman from 1999–2002.

Langer has received over 220 major awards. He is one of four living individuals to have received both the U.S. National Medal of Science (2006) and the U.S. National Medal of Technology and Innovation (2011). He also received the 2002 Charles Stark Draper Prize, considered the equivalent of the Nobel Prize for engineers, the 2008 Millennium Prize, the world's largest technology prize, the 2012 Priestley Medal, the highest award of the American Chemical Society, the 2013 Wolf Prize in Chemistry, the 2014 Breakthrough Prize in Life Sciences, the 2014 Kyoto Prize, and the 2015 Queen

Elizabeth Prize for Engineering. He is the also the only engineer to receive the Gairdner Foundation International Award; 87 recipients of this award have subsequently received a Nobel Prize. Among numerous other awards Langer has received are the Dickson Prize for Science (2002), Heinz Award for Technology, Economy and Employment (2003), the Harvey Prize (2003), the John Fritz Award (2003) (given previously to inventors such as Thomas Edison and Orville Wright), the General Motors Kettering Prize for Cancer Research (2004), the Dan David Prize in Materials Science (2005), the Albany Medical Center Prize in Medicine and Biomedical Research (2005), the largest prize in the U.S. for medical research, induction into the National Inventors Hall of Fame (2006), the Max Planck Research Award (2008), the Prince of Asturias Award for Technical and Scientific Research (2008), the Warren Alpert Foundation Prize (2011) and the Terumo International Prize (2012). In 1998, he received the Lemelson-MIT prize, the world's largest prize for invention for being "one of history's most prolific inventors in medicine." In 1989 Langer was elected to the Institute of Medicine of the National Academy of Sciences, and in 1992 he was elected to both the National Academy of Engineering and to the National Academy of Sciences. He is one of very few people ever elected to all three United States National Academies and the youngest in history (at age 43) to ever receive this distinction. In 2012 he was elected to the National Academy of Inventors.

Forbes Magazine (1999) and Bio World (1990) have named Langer as one of the 25 most important individuals in biotechnology in the world. Discover Magazine (2002) named him as one of the 20 most important people in this area. Forbes Magazine (2002) selected Langer as one of the 15 innovators worldwide who will reinvent our future. Time Magazine and CNN (2001) named Langer as one of the 100 most important people in America and one of the 18 top people in science or medicine in America (America's Best). Parade Magazine (2004) selected Langer as one of 6 "Heroes whose research may save your life." Langer has received honorary doctorates from Harvard University, the Mt. Sinai School of Medicine, Yale University, Western University (Canada), the ETH (Switzerland), the Technion (Israel), the Hebrew University of Jerusalem (Israel), the Universite Catholique de Louvain (Belgium), Rensselaer Polytechnic Institute, Willamette University, the University of Liverpool (England), Bates College, the University of Nottingham (England), Albany Medical College, Pennsylvania State University, Northwestern University, Uppsala University (Sweden), Tel Aviv University (Israel), Boston University, Ben Gurion University (Israel), Drexel University, Hanyang University (South Korea), Memorial Sloan Kettering, Karolinska Institutet (Sweden), Hong Kong University of Science, Carnegie Mellon University, University of Maryland, University of Laval (Canada), University of Limerick (Ireland), University of Illinois, National Institute of Astrophysics, Optics, and Electronics (Mexico) and the University of California — San Francisco Medal. He received his Bachelor's Degree from Cornell University in 1970 and his Sc.D. from the Massachusetts Institute of Technology in 1974, both in Chemical Engineering.

The topic I will be discussing is "Synthetic Tissues". This research started when I was a postdoctoral fellow with Judah Folkman trying to isolate the first substance that could regulate vascular growth. If one looks back at the history of biology and medicine, very often people isolate what are called "factors". One of the critical issues in isolating a factor, is having a bioassay. However, we didn't have any to study blood vessel growth. So we wanted to invent one. What we talked about was one way to study vascular growth was in the rabbit eye where you could visualize the blood vessels. If you put a tumor in the cornea it could secrete a substance that would make blood vessels grow; we could follow the way the blood vessels grew with an ophthalmic microscope. As I will go over, it would end up taking many months for the blood vessels to grow. In a paper, we wrote in *Nature* in 1976, we showed that you could have a polymer that could release the substances that we'd isolated and see if blood vessels grew[1]. So those were the bioassays that we developed to study blood vessel growth.

We also, wanted to find substances that might stop blood vessels from growing. As a source for such substances, one of the things we thought about was cartilage which is in your nose or your knee. So, I started out getting cartilage from small animals in the lab, but I couldn't get much cartilage from them so I went to a slaughterhouse in Cambridge and I got some cow bones. I probably spent about 40 or 50 hours a week scraping meat off the bones. Then I'd slice cartilage off and extract it in solvents like guanidine hydrochloride. Then I put the dialyzed extracts through different purification columns. At the end of several years, I'd have all these different, "fractions" that we wanted to test. Now, the challenge was a lot of the molecules in cartilage were ionic and some of them were large. Large molecules have very short lifetimes when you put them into the body; many are 25 minutes or less. However, the assays took many months. So, what we thought for this cornea assay that I mentioned, is we really need a polymer that would be inert in the eye, which is very difficult; but probably even more difficult is to be able to release molecules of any size for a long time.

After a number of years, against the conventional wisdom in the polymer community, we did create an approach that enabled biocompatible polymers to release molecules of any size for over 100 days. Since we could do this, we could now study angiogenesis. So, first, let's see if we can make blood vessels. That's one of the tissues. Here is a polymer in the rabbit eye (Figure 1) and now we put a

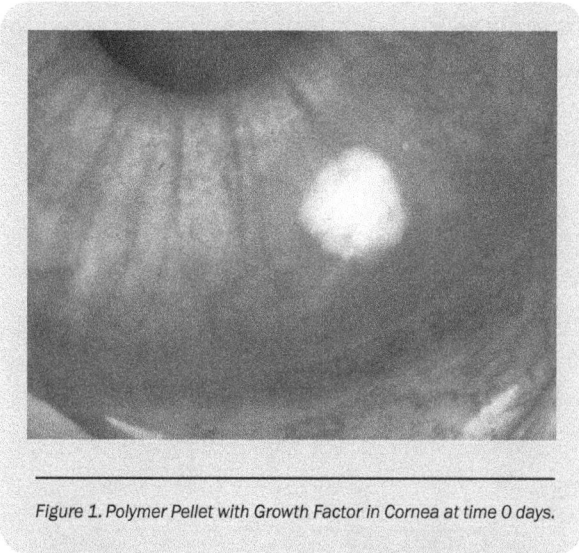

Figure 1. Polymer Pellet with Growth Factor in Cornea at time 0 days.

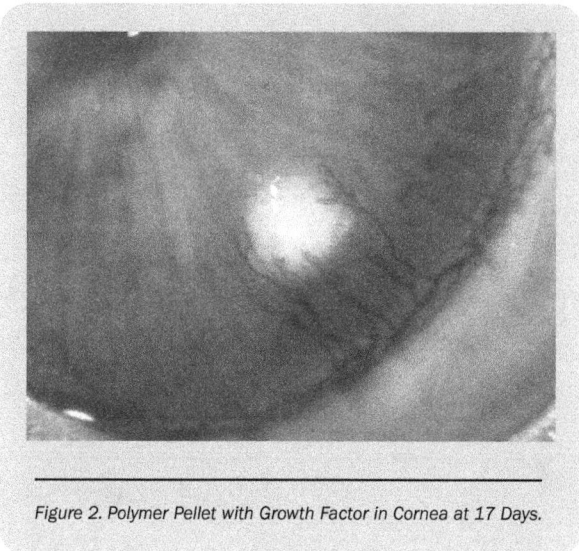

Figure 2. Polymer Pellet with Growth Factor in Cornea at 17 Days.

growth factor in it. What we saw is you can make brand new blood vessels; they grow right to the polymer (Figure 2). If you had the inhibitor, (this is from a paper we wrote in *Science* in 1976), here's the tumor, here's the polymer (Figure 3). What you see is 10 weeks later, if you don't have this cartilage derived inhibitor (CDI), you see a sheet of blood vessels growing over the polymer. But when you do have the polymer with the inhibitor, the blood vessels are much sparser, they grow more slowly and they actually avoid the polymer (Figure 4). 60 percent of the time, these tumors did not grow[2].

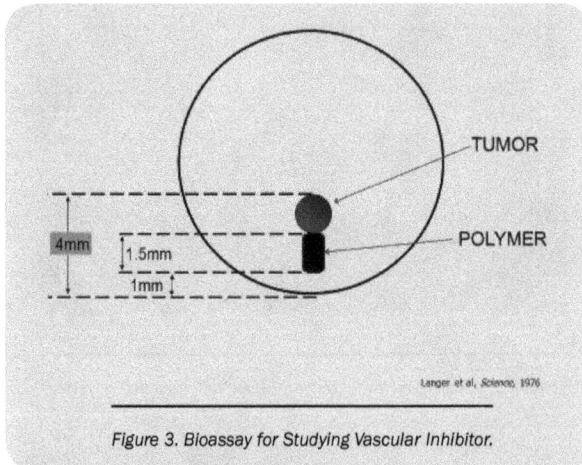

Langer et al, *Science*, 1976

Figure 3. Bioassay for Studying Vascular Inhibitor.

Rabbit corneal pocket assay

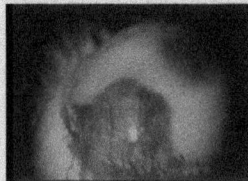

-CDI

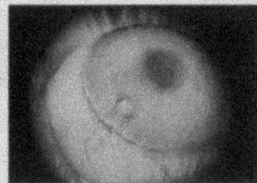

+CDI

Figure 4. Rabbit Cornea with and without Polymer Pellets Containing Cartilage Derived Inhibitor (CDI) at Week 10.

This was in 1976. It took another 28 years from the study we wrote in *Science*[2] before the first blood vessel inhibitors were approved by FDA. It took billions of dollars' and great work by companies like Genentech. Today, Vascular Inhibitors been approved for many types of cancer, and eye diseases like macular degeneration and diabetic retinopathy (Table 1).

Table 1. Angiogenesis Inhibitors Approved for Clinical Use *(Partial List)*.

Date Approved	Drug	Disease
February 2004	Avastin (Bevacizumab)	Colorectal Cancer
November 2004	Tarceva (Eriotinib)	Lung Cancer
December 2004	Macugen	Macular Degeneration
December 2005	Nexavar (Sorafenib)	Kidney Cancer
December 2005	Revimid	Myelodysplastic Syndrome
January 2006	Sutent (Sunitinib)	Gastric (GIST), Kidney Cancer
June 2006	Lucentis	Macular Degeneration
May 2007	Torisel (CCI-779)	Kidney Cancer
November 2007	Nexavar (Sorafenib)	Hepatocellular Carcinoma
February 2008	Avastin	Breast Cancer
May 2009	Avastin	Glioblastoma
November 2010	Afinitor	Giant Cell Astrocytoma
April 2011	Zactima (Vandetanib)	Medullary Thyroid Cancer
May 2011	Sutent	Pancreatic Neuroendocrine Tumors
November 2011	Eylea (Afibercept)	Macular Degeneration
January 2012	Axitinib (AG-013736)	Kidney Cancer
July 2012	Afinitor	Breast Cancer
September 2012	Eylea (Afibercept)	Central Retinal Vein Occlusion
January 2013	Avastin	Metastatic Colorectal Cancer
February 2013	Pomalyst (Pomalidomide)	Multiple Myeloma
April 2014	Cyramza	Advanced Stomach Cancer
August 2014	Avastin (Bevacizumab)	Cervical Cancer
November 2014	Avastin	Recurrent Ovarian Cancer
December 2014	Cyramza (Ramucirumab)	Non-small Cell Lung Cancer
February 2015	Lucentis	Diabetic Retinopathy with DME
February 2015	Lenvima (Lenvatinib)	Thyroid Cancer
April 2017	Lucentis	Diabetic Retinopathy

In addition, since I was doing work in materials, I was curious how did everybody decide what materials to use in medicine. So, I started to look into this while I was at the hospital. It was really interesting. What I saw almost exclusively is every time somebody used a material in medicine, it was a medical doctor who was really driving this. What they did because they urgently wanted to solve a problem, is they usually went to their house and they would find some object in their house that resembled the organ or tissue they wanted to fix. Here's just a few examples (Table 2).

Table 2. Examples of Polymers in Medicine.

Medical Use	Initial Use	Polymer
Artificial Heart	Ladies Girdles	Polyether Urethane
Dialysis Tubing	Sausage Casing	Cellulose Acetate
Vascular Graft	Clothing	Dacron
Breast Implants	Lubricant	Silicone
	Mattress Stuffing	Polyurethane

So, I'll start with the artificial heart. In the case of the artificial heart, you go back to 1967 and people at the NIH wanted to make an artificial heart and they said, what object has physical properties like a heart. They said a lady's girdle. It's got a good flex life. So they took the material in the lady's girdle and they made the artificial heart out of it. That's still true today 50 years later. One of the problems that occurs when you use this material is when blood hits the surface of the artificial heart, (the lady's girdle material), it can form a clot and that clot can go to the patient's brain, and they get a stroke and they die. But if you think about it, something that was designed to be a lady's girdle is probably not the optimal blood contacting material. This problem pervades all of medicine. One other example is breast implants. One of the materials used in a breast implant was actually a mattress stuffing. Being a chemical engineer, I thought, maybe we can do better. One of the things that you learn in chemical engineering is design. What our lab decided to do is design bio-materials on the blackboard from an engineering, chemistry and biology standpoint, and then synthesized them. Another idea, that Jay Vacanti who was a surgeon, and I started talking about was could we literally make new tissues and organs from scratch. Our idea was based on the observation that if you take the cells and put them into the body, they don't do much, but the cells are smart. If you put cells close enough together, they can actually reorganize themselves. In fact a group at Berkeley has shown that you can take mammary epithelial cells, put them close enough together, and they can make acinae and make milk. What Jay and I thought about is could we make a polymer template for cells that would help them reorganize. Also, the other thing that we realized is that pretty much everybody else was growing cells on polymers or tissue culture plastic in two dimensions. But in real life, so many things are in three dimensions. If you look at your skeletal system or anything else, usually it's in three dimensions. So we decided what we really need to do is make these scaffolds in three dimensions. We also needed to figure out the right cues. Sometimes to make the right tissue or organ, you have to put the right stresses on the cells

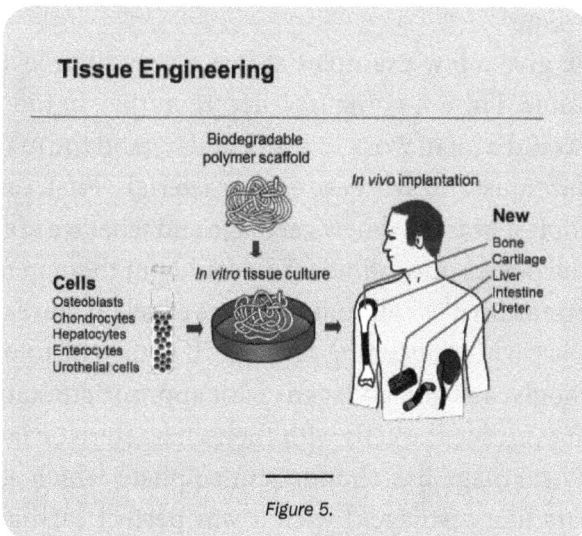

Figure 5.

to get the cells to do the right things. So, each of these steps ends up being important.

So, just to go over a couple things in creating a tissue (Figure 5). One of the things is the polymer. The polymers that sometimes we use are lactic-glycolic acid copolymers (PLGA). These have a nice feature that they'll break down into water and carbon dioxide, but they don't allow you to get cell attachment for certain cell types. So we also synthesized new polymers that looked a lot like PLGA, but we'd add a lysine group[3]. Then through a carbodiimide reaction, you could add almost anything you want so you could really control the surface by perhaps attaching specific amino acids. Then, what you might do is convert these into a physically branched structure. We also thought about using CAD/CAM techniques (computer-aided design) to make different shaped systems and then use techniques like three-dimensional printing or foaming techniques to make them any shape you want. So just to speculate 40 or 50 years from now, let's say somebody comes to a plastic surgeon and they don't like their nose and they want a new nose. Well, what we thought about is we could make a nose scaffold with CAD/CAM techniques. Then you could take their own cells, put them on the scaffold and make the tissue[4].

Let me give a few examples at just a very high level to show what's possible. I'm not going into depth on this, but in each case, our students did a lot of work to really understand the materials, the reactor conditions, and synthesize new materials. First, cartilage—A million people a year need new cartilage and what we are able to do is put chondrocytes on scaffolds. You could put them in nude mice. We redid one animal's skull, and another's cheek. If you examine the tissue, it's pure white glistening cartilage. If you do histology you see glycosaminoglycans and collagens with appropriate stains (Figure 6). You can even see the cells with their characteristic lacunae and then finally cartilage has chondroitin sulphate which stains blue. So it appears like cartilage. I wish it was perfect cartilage. It isn't. Actually if anybody is thinking that they are a runner and we can

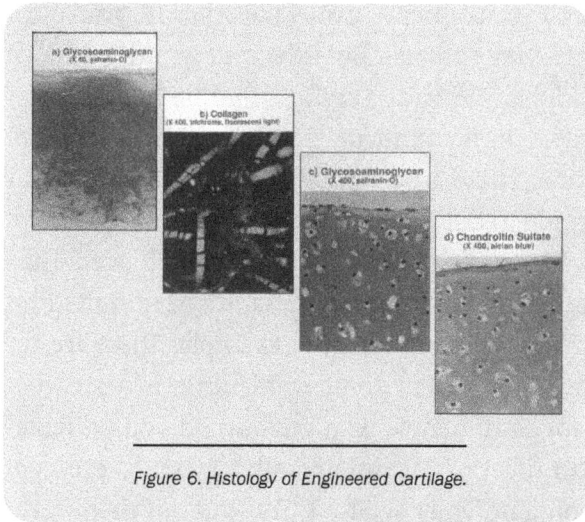

Figure 6. Histology of Engineered Cartilage.

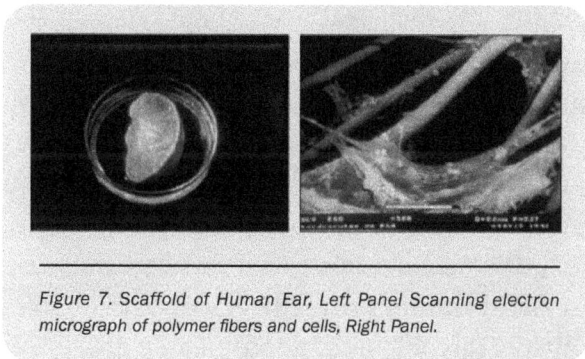

Figure 7. Scaffold of Human Ear, Left Panel Scanning electron micrograph of polymer fibers and cells, Right Panel.

help them, we're not there yet. We're working on it. But what you can do is really help people, in some cases who have different cosmetic issues.

So, the army funded myself, Linda Griffith, who was one of my former post-docs and now a professor at MIT and Jay Vacanti. They were concerned that soldiers that come back from Iraq and Afghanistan sometimes don't have body parts like ears. Could we make that? Linda made an ear. There are fibers with cartilage cells (Figure 7). The cells keep multiplying. Ultimately the polymer totally dissolves and you get the ear. We haven't done human ears yet, but we have done some patients. We helped a 12-year-old boy that my colleague Jay Vacanti operated on. Like other 12-year-old boys in the US he likes to play baseball. But if he ever got hit in the chest with a ball, he could die. We made a scaffold for him, took his own cells and Jay operated on him and made him a new chest. He's now an adult and doing fine. There are a number of examples like this where you can help people in different situations.

Skin is the next example that I want to go over. There are now different companies developing these. It's been quite dramatic. I'm just going to discuss a very positive example. These are actually now FDA approved. What I'm going to mention is what we and others have done for both people who are burned and patients who have diabetic skin ulcers. You can actually cryopreserve neonatal skin fibroblasts on a polymer scaffold. You can put them on a person at the time of injury and they heal. The patient gets new skin.

It also occurred to us that there might be a variety of applications for these technologies. They could also be very useful for new drug testing, for toxicity, and things like that, like an organ on a chip. I'll just give you an example of that. This was done by Gordana Vunjak, who was in our lab at the time. Now she is a professor at Columbia. So what we did is take cardiomyocytes (heart muscle cells), put them on a scaffold and they made the characteristic ultrastructure for cardiac tissue, sarcomeres and intercalated disks. You can have six million heart cells beating at once. Then you can actually take an EKG of it. You can put different drugs on it to see how they'll behave on cardiac tissue. Many drug companies are now using this for new screening techniques to see if their drugs are safe.

The next example is spinal cord. Erin Lavik was one of our graduate students and she wondered whether we could help people who were paralyzed. So what she did is design a scaffold that would mimic the gray and white matter of the spinal cord. There would be an outer part that might be nano-patterned that could help provide axonal guidance, and an inner part which was porous and could be seeded with neuronal stem cells. Here we collaborated with Ted Tang, who is a neurosurgeon, and Evan Snyder who is a neural stem cell expert. In the experimental model that Ted and Evan developed, you make the rats paraplegic by taking out a four millimeter hemisection of the spinal cord. Then we did four sets of animals.

1. Control – a sham operation.
2. Cells by themselves.
3. Polymer by itself.
4. Polymer – cell system.

The idea is that when you have a gap from a contusion you put the scaffold in it – it acts as a bridge and you get fewer cysts, you spare white matter, and you even get some remodeled tissue. In particular, we looked for what happens in terms of function. Can the rats do better? We did 12–13 animals in each of the 4 groups. If you do nothing or add cells by themselves, the animals can't bear their own weight, and the paws are splayed in an awkward fashion. There is a BBB scoring system. 20 is what a regular animal would get; these controls get a 5. When you put the cells in by themselves you also get a 5. Interestingly, the scaffold by itself gets about an 11 and the scaffold and cells gets a 14. The animals can bear their own weight, and the paws are splayed normally[5]. Then these experiments were repeated in non-human primates, and getting almost the same results[6]. Then the company that licensed this technology (*In vivo therapeutics*) got FDA approval to start clinical trials. They've now got institutional review board approval at 33 hospitals. Basically, if somebody gets paralyzed, within the first week after the injury you image it with MRI and slip the scaffold in. The company decided to do the clinical trial without the stem cells just because from a regulatory standpoint it's simpler. But it's been very exciting. It's really early, but let me tell you about the first patient.

A 25-year-old boy was in a motorcycle accident in Arizona and they operated on him and he showed remarkable improvement three months later. There were no adverse events. He has active movements of his hip flexors, and palpable contractions of knee extensors. He had no bowel function initially, but regained it.

He had no bladder function and that's improved. In general, the percentage of patients who are paralyzed and improve is about 15 percent, but here so far it's 62.5 percent based on an ASIA scoring system. Again, it's too early to know what'll happen for sure but it's been encouraging to see that tissue engineering might help.

The last example I want to talk about is the pancreas. This is a problem that the Juvenile Diabetes Foundation called me up about 10 years ago to see whether we might be able to help because of what we've done in materials. In particular, there's a strategy that people have used for diabetes. I should point out, that for all the other systems that I talked about, there could be rejection. The reason there's not rejection is we might use the patient's own cells in some cases, we might use no cells in some cases, or we might give immunosuppressive drugs. That's actually what they do for transplants now. But one of the other things that we thought about is could we come up with approaches and materials where you would not worry about any of that. So, if you had a pancreas, one of the things that you might do, is put what are called beta cells (Insulin producing cells) inside a membrane. If you have the membrane permeability parameters just right, glucose which is a small molecule and insulin which is a little bigger might diffuse back and forth through the membrane. But entities' that cause rejection like immune cells or antibodies, are much bigger and they can't get through the membrane. So, the idea is could we make a device that could have just the right porous structure but protect Beta cells from the immune system. People have worked on this for a long time. One of the materials people use is alginate. That's actually from seaweed. Almost all the time when you do microencapsulation with polymers, organic solvents are required for encapsulation. That might work for encapsulated molecules, but would destroy cells. Cells are much more fragile. The nice thing about alginate though, is that it has all these charged groups. So, if you just inject it into a water bath of a calcium or barium salt, it'll immediately form a bead and the cells will be encapsulated. It's just done in water. So it's very, very mild.

That would be great except for one thing. Alginate is not sufficiently biocompatible and the body recognizes it as foreign. This issue has stopped progress. What the Juvenile Diabetes Foundation asked is, is there some way that we could create materials to keep cells alive and functioning and not get covered with scar tissue. This is what we call super biocompatible materials. So Dan Anderson, who was one of my post-docs, and is now a professor at MIT and I thought, was maybe one way to address this problem is to develop novel high throughput chemistry approaches. One of the biggest problems, of course, is if you have tissue, how do you evaluate it *in vivo*. What people normally do when they put any implant in a person is the way they evaluate it is to take it out, and then they might look at it optically or by histology. But that's incredibly slow. We wanted to be able to speed things up. So we developed a mouse model where we could do it faster. Sometimes people call this a high throughput mouse. I think at best it's an intermediate throughput mouse. What we would do is we would inject different capsules into the animal. If there's going to be inflammation, usually you would get reactive oxygen species. However, we thought we could inject a dye that might react with those species like luminol and if it lights up, that tells you pretty quickly that you, at least, as a screen might have inflammation[7]. So, we used this as one tool for speeding up our studies. We also found interestingly that the size of the capsules ended up making a big difference. As you make the microcapsules bigger and bigger, the response goes away. For example, you get less and less tissue deposition as the implanted microcapsules get bigger and bigger[8]. Then we synthesized thousands of alginate derivatives by a combination of robotics and different chemistries[9]. We would basically start with alginate and then we would do various derivatizations to come up with novel material designs. We characterized the materials extensively. I should point out we've purified them extensively too and there are many different controls that have been done to look at different other possibilities that might cause a problem. Then we put the capsules in different animal models and even in different locations, and then in diabetic rats and non-human primates. If you use regular alginate, in non-human primates, you actually get very nice capsules. But what happens is, you get fibrotic overgrowth, you get a lot of blood vessel growth, and the islets are dead by four weeks. However, when you use the best of our materials what happens is you get intact capsules and healthy islets, and no fibrosis[9]. Now, we've actually gone out to six months, and get good viability. Even at six months, you see almost no encapsulation.

Finally, just to go over the problem we're trying to solve, you take diabetic animals and you take rat islets and put it in these capsules, and you can control blood glucose for over 200 days (Figure 8). In a collaboration with Doug Melton, who has isolated human islet cells, you get control of blood glucose for at least 175 days[10]. In summary, I'Ve discussed how chemistry and engineering can be used to create synthetic tissues. There is a long way to go but I believe advances will continue to be made in chemistry, biology, and engineering in this area and that this field will have an enormous impact on human health and our well-being.

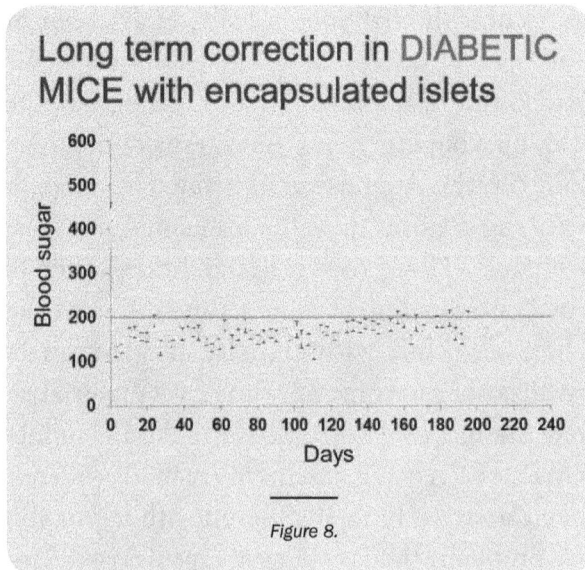

Figure 8.

References

1. Langer, R. & Folkman, J. Polymers for the sustained release of proteins and other macromolecules. *Nature* **263**, 797–800 (1976).
2. Langer, R., Brem, H., Falterman, K., Klein, M. & Folkman, J. Isolation of a cartilage factor that inhibits tumor neovascularization. *Science* **193**, 70–72 (1976).
3. Barrera, D., Zylstra, E., Lansbury, P. & Langer, R. Synthesis and RGD peptide modification of a new biodegradable copolymer system: Poly(lactic acid-co-lysine). *J. Am. Chem. Soc.* **115**, 11010–11011 (1993).
4. Shastri, V., Martin, I. & Langer, R. Macroporous polymer foams by hydrocarbon templating. *Proc. Natl. Acad. Sci.* **97**, 1970–1975 (2000).
5. Teng, Y.D., Lavik, E.B., Qu, X., Park, K., Ourednik, J., Zurakowsi, D., Langer, R. & Snyder, E.Y. Functional recovery following traumatic spinal cord injury mediated by a unique polymer scaffold seeded with neural stem cells. *Proc. Natl. Acad. Sci. USA* **99**, 3024–3029, 2002.
6. Slotkin, J., Pritchard, C., Luque, B., Ye, J., Layer, R., Lawrence, M., O'Shea, T., Roy, R., Zhong, H., Vollenweider, I., Edgerton, V., Courtine, G., Woodard, E. & Langer, R. Biodegradable scaffolds promote tissue remodeling and fuctional improvement in non-human primates with acute spinal cord injury. *Biomaterials* **123**, 63–76 (2017).
7. Bratlie, K., Dang, T., Lyle, S., Nahrendorf, M., Weissleder, R., Langer, R. & Anderson, D. Rapid biocompatibility analysis of materials via *in vivo* fluorescence imaging of inflammatory responses. *PLoS ONE* **5**, 1–8 (2010), PMID: 20386609, PMCID: PMC2850367.
8. Veiseh, O., Doloff, J., Ma, M., Vegas, A., Tam, H., Bader, A., Li, J., Langan, E., Wyckoff, J., Loo, W., Jhunjhunwala, S., Chiu, A., Siebert, S., Tang, K., Hollister-Lock, H., Aresta-Dasilva, S., Bochenek, M., Mendoza-Elias, J., Wang, Y., Qi, M., Lavin, D., Dholakia, N., Thakrar, R., Wier, G., Oberholzer, J., Greiner, D., Langer, R. & Anderson, D. Size and shape dependent foreign body immune response to materials implanted in rodents and non-human primates. *Nat. Mater.* **14**, 643–651 (2015), PMCID: PMC4477281.
9. Vegas, A., Veiseh, O., Doloff, J., Ma, M., Tam, H., Bratlie, K., Li, J., Bader, A., Langan, E., Olejnik, K., Fenton, P., Kang, J., Hollister-Locke, J., Bochenek, M., Chiu, A., Siebert, S., Tang, K., Jhunjhunwala, S., Aresta-Dasilva, S., Dholakia, N., Thakrar, R., Vietti, T., Chen, M., Cohen, J., Siniakowicz, K., Qi, M., McGarrigle, J., Lyle, S., Harlan, D., Greiner, D., Oberholzer, J., Weir, G., Langer, R. & Anderson, D. Combinatorial development of hydrogels that mitigate the

foreign body response in primates. *Nat. Biotechnol.* **34**, 345–352 (2016), PMCID: PMC4904301.

10. Vegas, A., Veiseh, O., Gurtler, M., Millman, J., Pagliuca, F., Bader, A., Doloff, J., Li, J., Chen, M., Olejnik, K., Tam, H., Jhunjhunwala, S., Langan, E., Aresta-Dasilva, S., Gandham, S., McGarrigle, J., Bochenek, M., Hollister-Lock, J., Oberholzer, J., Greiner, D., Weir, G., Melton, D., Langer, R. & Anderson, D. Long-term glycemic control using polymer encapsulated human stem cell-derived beta cells in immune-competent mice. *Nat. Med.* **22**, 306–311 (2016), PMCID: PMC4825868.

Dynamics of Molecular Processes and the Sustainable Development of Humanity*

Yuan Tseh Lee[1]

Yuan Tseh Lee was instrumental in the development and construction of an apparatus that utilized crossed molecular beams, presenting a breakthrough technique that allowed for the understanding of the dynamics of elementary chemical reactions. This was done by following the trajectories of reactants and reaction products in single collision events, allowing the visualization of the dynamics of how chemical reactions take place. This article also highlights Prof. Lee's belief in the severity of the consequences of global warming and his concerns relating to the need to substantially reduce carbon emissions.

Keywords: Dynamics; Molecular; Sustainable; Development; Humanity.

---

* This chapter was based on a lecture delivered by Professor Lee at the Molecular Frontiers Symposium in the Tokyo University of Science, Kagurazaka campus in October 2016.
[1] Nobel Laureate in Chemistry (1986). Institute of Atomic and Molecular Sciences, Academia Sinica, Taiwan.

About Yuan Tseh Lee

Yuan Tseh Lee (born 19 November 1936) is a Taiwanese chemist. He was the first Taiwanese Nobel Prize laureate, who, along with the Hungarian-Canadian John C. Polanyi and American Dudley R. Herschbach, won the Nobel Prize in Chemistry in 1986 "for their contributions to the dynamics of chemical elementary processes".

Lee's particular physical chemistry work was related to the use of advanced chemical kinetics techniques to investigate and manipulate the behavior of chemical reactions using crossed molecular beams. From 15 January 1994 to 19 October 2006, Lee served as the President of the Academia Sinica of Taiwan. In 2011, he was elected head of the International Council for Science.

In February 1967, he started working with Dudley Herschbach at Harvard University on reactions between hydrogen atoms and diatomic alkali molecules and the construction of a universal crossed molecular beams apparatus. After the postdoctoral year with Herschbach he joined the University of Chicago faculty in 1968. In 1974, he returned to Berkeley as professor of chemistry and principal investigator at the Lawrence Berkeley National Laboratory, becoming a U.S. citizen the same year. Lee is a University Professor Emeritus of the University of California system.

One of the major goals of chemistry is the study of material transformations where chemical kinetics plays an important role. Scientists during the 19th century stated macroscopic chemical processes consist of many elementary chemical reactions that are themselves simply a series of encounters between atomic or molecular species. In order to understand the time dependence of chemical reactions, chemical kineticists have traditionally focused on sorting out all of the elementary chemical reactions involved in a macroscopic chemical process and determining their respective rates.

Swedish chemist Svante Arrhenius studied this phenomenon during the late 1880s, and stated the relations between reactive molecular encounters and rates of reactions (formulated in terms of activation energies).

Other scientists at the time also stated a chemical reaction is fundamentally a mechanical event, involving the rearrangement of atoms and molecules during a collision. Although these initial theoretical studies were only qualitative, they heralded a new era in the field of chemical kinetics; allowing the prediction of the dynamical course of a chemical reaction.

In the 1950s, 1960s and 1970s, with the development of many sophisticated experimental techniques, it became possible to study the dynamics of elementary chemical reactions in the laboratory. Such as the analysis of the threshold operating conditions of a chemical laser or the spectra obtained using various linear or non-linear laser spectroscopic techniques.

Lee's research focused on the possibility to control the energies of the reagents, and to understand the dependence of chemical reactivity on molecular orientation, among other studies related to the nature of reaction intermediates, decay dynamics, and identifying complex reaction mechanisms. To do so, Lee used a breakthrough laboratory technique at the time, called the *"crossed molecular beams technique"*, where the information derived from the measurements of angular and velocity distributions allowed him and his team to understand the dynamics of elementary chemical reactions.

In addition to the Nobel Prize, his awards and distinctions include Sloan Fellow (1969); Fellow of American Academy of Arts and Sciences (1975); Fellow of American Physical Society (1976); Guggenheim Fellow (1977); Member National Academy of Sciences (1979); Member International Academy of Science, Member Academia Sinica (1980); E.O. Lawrence Award (1981); Miller Professor, Berkeley (1981); Fairchild Distinguished Scholar (1983); Harrison Howe Award (1983); Peter Debye Award (1986); National Medal of Science (1986). Lee was awarded the Othmer Gold Medal in 2008 in recognition of his outstanding contributions to progress in chemistry and science.

His post-doctoral supervisor and fellow Nobel Laureate Dudley Herschbach called Lee "The Mozart of physical chemistry".

I like baseball. My interest in baseball came about just after the Second World War, when I was in the third grade. At this time we were told that if you are good enough to make the baseball team that you would not have to attend the classes in the afternoon. This inspired me as I disliked going to class and as a result I became a very good second baseman. Consider the hitting of the baseball; this action resembles the scattering between two hard spheres. So the impact parameter, namely the aiming error, will determine whether you hit a home run, a pop-up or a ground ball. The aiming error is especially important as to the ball's trajectory. But if there is a force operating in between them, this will make things a little more complicated. For example if a molecule moves towards another molecule that has a repulsive force, it would be bounced backward. On the other hand, if the interaction is an attractive one then the molecule would bend over. So in one sense repulsion excludes the volume and attraction seems to expand the volume. And in the real world you are looking at the consequences of the interaction between atoms, or atoms and molecules. At a long distance, because of the dispersion force which is caused by the fluctuation of electrons density around the nuclei, the instantaneous dipoles created make the interaction between atom and molecule attractive. On the other hand, when the electron cloud moves close together it would become repulsive. So in the real world, if you look at the atom–molecule interaction, you will always find that the interaction is attractive at a long distance, and repulsive at a short distance. Then when you look at the scattering process between two atoms it becomes really interesting. If the impact parameter is zero it will bounce back, if you increase the aiming error it will become a line-drive and a home run and then gradually become a pop-up. But if you continue to increase the aiming error and observe the scattering angle, because of the attractive force becoming more important than the repulsive force, the scattering angle will move from positive to zero and then to negative, and after reaching a maximum value, it will move back to zero. This is what is happening in real atomic collisions.

And that is really important. If you look at thermal-conductivity, viscosity of gaseous species and observe how they deviate from an ideal gas case, all these depend on the detail of scattering between atoms and molecules. These in fact depend on the interaction forces between atoms and molecules.

If you look at the chemical reaction, we know that the collision will bring about the chemical change. So if you mix hydrogen molecules and chlorine molecules, it will transform into hydrogen chloride ($H_2 + CL_2 \rightarrow 2HCL$). But as you know this equation only shows the stoichiometric relation between molecules, that of one molecule and one chlorine molecule being transformed into two hydrogen chlorine (HCL). However if you really mix hydrogen and chlorine together in the dark, although there are lots of collisions between H_2 and Cl_2, chemical change might not take place because it takes a very high collision energy to overcome the barrier to induce this change. On the other hand, if you shine light into the mixture, chlorine will dissociate and a new molecule, hydrogen chloride (HCl) will be produced. Sometimes hydrogen chloride produced could be cold, sometimes it could be warm. If you follow the trajectories of the atoms and molecules, what you might see is the chlorine molecule dissociated into atoms and then the chlorine atom colliding with a hydrogen molecule to produce a hydrogen chloride and a hydrogen atom. This is the second step. The hydrogen atom produced will then react with a chlorine molecule and form HCl and Cl. So HCl is produced in two ways, via the collision of the chlorine atom with hydrogen molecule, or the hydrogen atom with chlorine molecules as shown in Fig. 1.

So for most of us learning about chemical kinetics, we try to understand the mechanism and the specific rate constant of each step. These determine how a system moves systematically from reactant to product as a function of time. For humankind one of the most important chemical reactions is the combustion of hydrocarbon. This simple chemical reaction provides more than

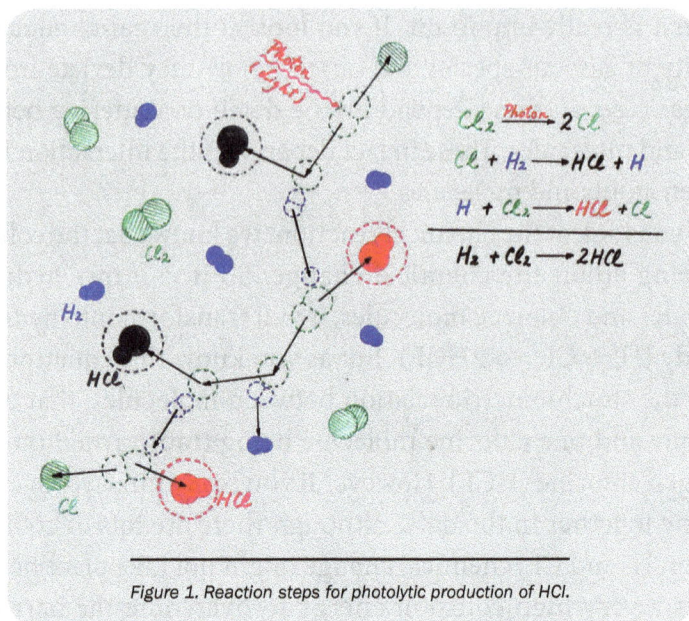

Figure 1. Reaction steps for photolytic production of HCl.

80% of today's energy requirements. Of course, the oxygen-oxygen bond and carbon-hydrogen bond are weaker than the carbon-oxygen bond in CO_2, and in the water molecule. That is why the transformation between methane and oxygen produce, carbon dioxide and water will produce a large amount of energy. As I said molecular collision induces chemical change, but again nothing will happen when you mix methane and oxygen unless you initiate the reaction. If you used a match to initiate the combustion process you will immediately see a bond rupture, addition of oxygen atom, dimerization, further elimination of hydrogen atoms, and step-by-step, carbon dioxide and the water molecule will be formed as end products. One of the final important steps is the reaction between carbon monoxide and hydroxyl radical (OH) to produce CO_2 + H.

The collision between carbon monoxide and a hydroxyl radical will form carbon dioxide and a hydrogen atom. This chemical reaction is quite interesting. The alignment of the molecule is important. However not only alignment is important, you also have to have enough energy to overcome the barrier. Otherwise, a chemical reaction will not take place as shown in Fig. 2. Over

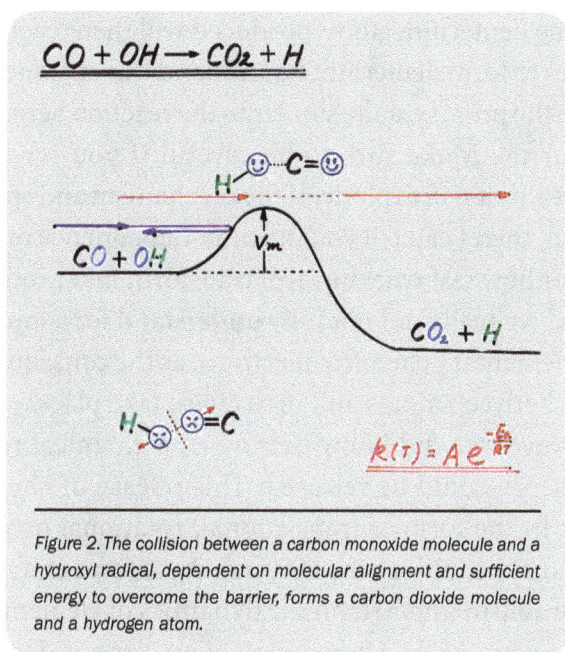

Figure 2. The collision between a carbon monoxide molecule and a hydroxyl radical, dependent on molecular alignment and sufficient energy to overcome the barrier, forms a carbon dioxide molecule and a hydrogen atom.

a hundred years ago, Arrhenius pointed out that the specific rate constant was dependent on temperature. In the Arrhenius equation ($K = Ae^{-\varepsilon/kt}$), the pre-exponential factor, or the A factor is actually determined by the probability of proper alignment and the activation energy (ε) which determines that temperature dependence is related to the potential energy barrier. Arrhenius also pointed out over a century ago, that if man continued to burn fossil fuels that produce carbon dioxide, the surface temperature of the world would rise. But I don't think at that time he was worried about the consequences of global warming.

Now I want to take you through my work on molecular beam scattering experiments. As I pointed out earlier, if you look at the chemical reaction, $F_2 + D_2 \rightarrow 2DF$ as an example you are only observing the stoichiometric relation. The first thing that you need to understand is the mechanism. In this case, fluorine molecules represent a very weak bond and so it would dissociate into atoms quite easily. The fluorine atom would then react with deuterium molecules to form deuterium fluoride molecules, plus deuterium atoms. The deuterium atoms produced will then react with fluorine molecules to form deuterium fluoride and this chain reaction would complete the process, quite similar to the reaction between hydrogen and chlorine. Make sure to be careful if you try this reaction yourselves, as it will explode. If the mechanism and specific rate are all known, then I can tell you the time-dependence of the chemical change, or how fast reactants will transform into products. The one thing that we really did not fully understand for a long time is how exactly elementary chemical reactions, as the consequence of single collision between atoms and molecules, take place.

However we did know that when a chemical reaction takes place, energy could be released. This release of chemical energy could be in the form of translational, rotational or vibrational. If our naked eye could see the movement of atoms and molecules, we would be able to understand the dynamics of elementary chemical reactions very easily. Unfortunately we cannot. However, in the laboratory it is possible to track the trajectories of atoms and molecules as shown in Fig. 3.

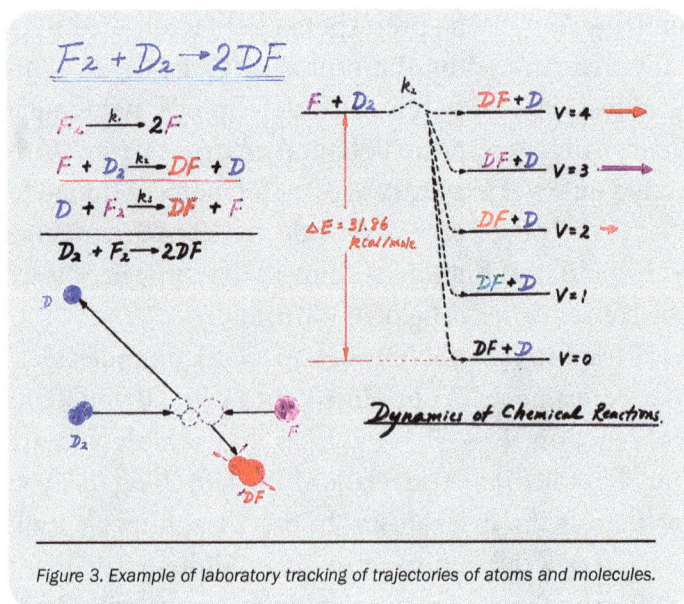

Figure 3. Example of laboratory tracking of trajectories of atoms and molecules.

So about 50 years ago, when I was doing my post-doctoral research with Professor Herschbach at Harvard, scientists in Germany, England, The United States, all around the world, were already working on trying to observe the scattering between atoms and molecules and witness the chemical change in a single collision. Yet no one had managed to do this, except for the scattering of alkali atoms. I was young and when I was told that something was impossible, it became a challenge that got me really excited. In 1967, we started to design an apparatus to try to carry out this experiment. Among the important discoveries that I made during this time related to the adage that you need to learn from your mistakes and that this is how to make progress. But what I learned in addition is that it is not only essential to learn from your own mistakes but it is also important to learn from the mistakes of others. Just to realize that someone made a mistake is not helpful. You really have to gain knowledge through the mistakes that others have made. So I began to try to understand why all the eminent scientists around the world had not been able to do this experiment and solve the problems. And in doing this I began to realize that everyone who had attempted this had been misled. The discussion relating to the problems with the experiment had centered mostly in the crossing region of molecular beams in a vacuum chamber, where the number of molecules produced was very small. From there, only about 1 million molecules per second will reach the detector for electron bombardment Ionization detection. The ionization efficiency is only one in ten thousand. As a result, the focus at the time was on addressing the issue of increasing ionization efficiency. This meant that everybody who was working on the perceived problem in a crossed molecular beam experiment was trying to design a machine that would improve the detection efficiency. But no one had realized that in order to make the experiment successful, it was not only necessary to have an improved signal level, but also a good signal to noise ratio. When these 1 million molecules passed through the detector, it was possible to produce about 100 ions. This is a large enough number to be detected with an ion counter, if there is not several million times this in background ions to overwhelm the signal. Our machine was specially designed with a three-part pumping mechanism so the partial pressure of the interfering molecules would be reduced by about one million times, compared to that in the scattering chamber. The cut-out view of our machine is shown in Fig. 4. We adopted the configuration of fixed beams and rotatable detector for the measurements of angular and velocity distributions of scattered molecules. Our efforts represented the first time that the angular and velocity distributions of scattered molecules

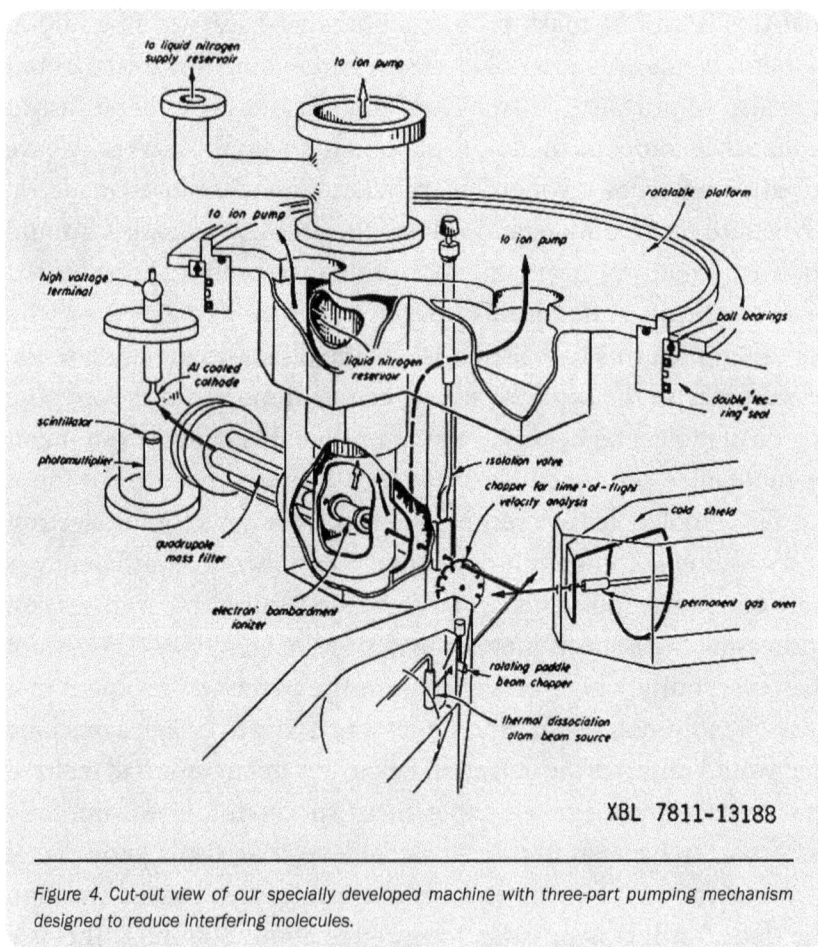

XBL 7811-13188

Figure 4. Cut-out view of our specially developed machine with three-part pumping mechanism designed to reduce interfering molecules.

of any species could be measured. When we published the paper on this, the scientific community was astonished. Until then it had been believed that it was almost impossible to even observe the scattering of rare gas atoms. However we did succeed in building the machine that could do it. If one can measure differential scattering cross sections of argon-argon scattering, one can derive the interaction potential, and from the interaction potential, you are able to calculate the macroscopic properties and transport phenomena of gaseous argon. As to the chemical reaction, the fluorine atom collided with the deuterium molecule and produced deuterium fluoride. When you scan the velocity and angular distributions that at low collision energy, as shown in Fig. 5, a fluorine atom and two deuterium atoms

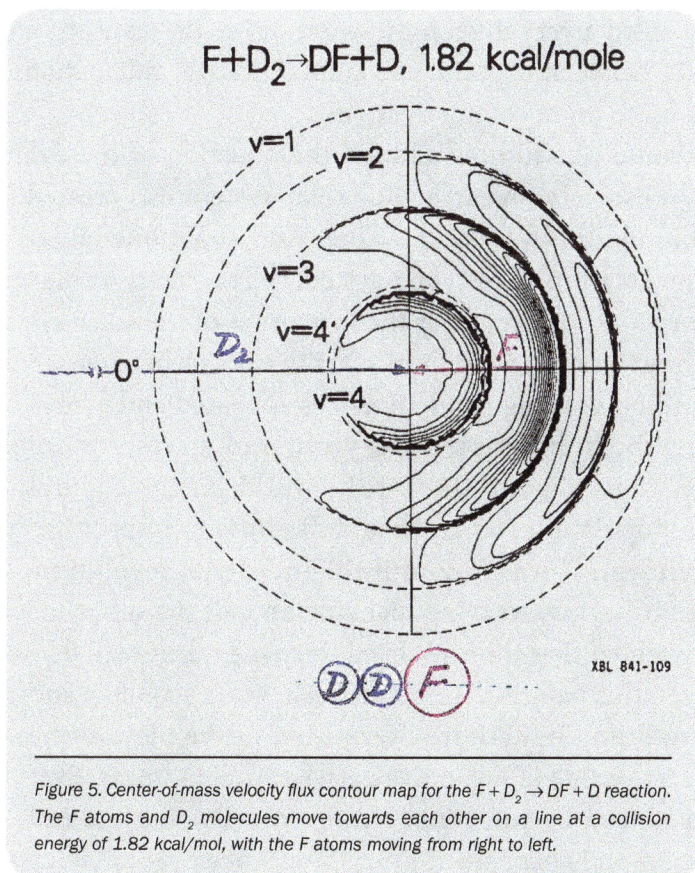

Figure 5. Center-of-mass velocity flux contour map for the $F + D_2 \rightarrow DF + D$ reaction. The F atoms and D_2 molecules move towards each other on a line at a collision energy of 1.82 kcal/mol, with the F atoms moving from right to left.

have to line up together, and the velocity distributions also indicate that most of the energy released is in the vibration of deuterium fluoride as the products move rather slowly.

Fifty years ago, when I worked on the helium scattering experiment, to determine the interaction potential, computers were not capable of complex theoretical calculations and could not come up with a reliable result. They could not even tell us whether helium dimer was in a bound state or not. Nowadays you don't even have to do the actual helium scattering experiment, you can just do the calculation to carry this out. This is true for even the reaction of a fluorine atom and hydrogen molecule. Today using supercomputers you can carry out the calculation and you will be able to tell what is going on. Certainly the present capabilities of supercomputing have enabled a very different way of finding the answers, and the interface between experiments and theoretical calculations have moved up to more complex systems.

I would like to now address the question of the evolution of human society on earth. Our solar system was created about 9.5 billion years after the big bang. And about one billion years after the creation of the solar system, phenomena of life started on the earth's surface. After the formation of the solar system, it took about another 4.5 billion years to get us where we are today. Two million years ago, our ancestors appeared and evolved. One species, the homo sapiens survived and started to dominate and began to migrate all over the world. At the beginning the human population was small and did little to impact the global environment. It was not until the industrial revolution, when the steam engine, weaving machine and all the other machines were invented that things started to change drastically. I want to show you this painting from about 300 years ago of an aboriginal tribal wedding in Taiwan. Everything in the picture shown on the left hand side of Fig. 6 was derived from biomass. Everything was produced by sunshine. However, following the industrial revolution, what happened? Everything that we see today, concrete,

Figure 6. Left panel: Taiwanese wedding scene ca 1700, materials are all "biomass" derived.
Right panel: Scene from the Industrial Revolution, shift towards materials dug from the ground.

steel, the fossil fuels used to produce these, are all dug out from the ground. Humankind used to be part of nature, however gradually we began to feel that we are the masters of nature. This dissociation and departure from nature is really causing a lot of problems. We know about the tremendous progress that took place in science and technology from 1850 until now. The alarming thing is that alongside these reported technological advancements, we are burning fossil fuels in higher and higher quantities in tangent with the population explosion.

At the beginning of the last century we had 1.5 billion people, by its end we had 6 billion people. The population is currently estimated at around 7.2 billion. By 2050 the United Nations projects that 9.7 billion people will inhabit the planet. And with this growth in population and consumption, we have clearly overloaded the earth and with this we are changing the global environment. This is the reality of the civilization that we live in now.

You have heard about free trade and globalization. All over the world, with the capital and technology flow across national borders without much hindrance, businesses are seeking out cheap labor sources for mass production. Countries like China, Vietnam and Indonesia provided cheap labor to produce consumer products for global consumption at the cost of using excessive natural resources. In the process we have destroyed the environment, created pollution and have caused CO_2 emissions to keep on rising. Globalization and free trade have enabled that production and consumption proceed at full speed ahead, inspired by significant increases in GDP, heralded as progress. Whether you like it or not this is what we are doing at present. It's a horrible thing.

We are exceeding our planetary boundaries and are in a very precarious position. If you look carefully we are living in the very thin layers on the surface of the earth that hover above the earth. These are the hydrosphere and atmosphere. We have been living within these thin layers for thousands of years. The planet's natural processes allowed for a stable climate. However, we are now changing our living environment. The first thing you will notice is the pollution, especially if you visit many major cities in Asia. Can you convince yourself that breathing a heavily polluted air is progress? We are suffocating ourselves. I do not believe that no matter how rich we become that this really can be considered progress.

The second thing which is the most important, is the energy balance. Sunshine mostly as visible photons provides the planet of earth a large amount of energy and the earth also emits energy as infrared photons to the universe. If there were no greenhouse gases, i.e. carbon dioxide, methane, the temperature on the planet would be about minus 20 degrees. Prior to the industrial revolution, the planet's carbon dioxide concentration was maintained at about 280 ppm for a long time, and at this level the global average temperature was stable at about 20 degrees. However now because greenhouse gas has increased we are trapping more energy than we are giving out, so that the temperature is rising, as seen in Fig. 7. Overall carbon dioxide concentration from the 1980s onwards has continued to go

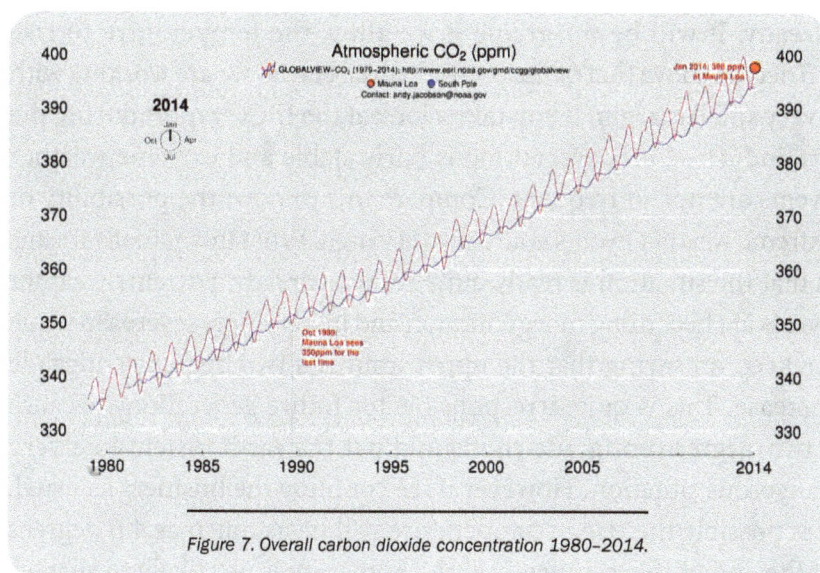

Figure 7. Overall carbon dioxide concentration 1980–2014.

up. By the year 2000 this was up to 370 ppm and it keeps going up and up. At present we are reaching concentration levels of 400 which are staggering and emissions are continuing to increase very fast. If you look at the period before the industrial revolution, the carbon dioxide concentration levels were maintaining at around 280 ppm. While there have been cyclical periods covering ice ages to warmer periods, these are due to the fact that the earth's orbit is not round and the axis of self-spinning makes an angle with the axis of the orbit. We are at the high point of the cyclic period and the surface temperature of the earth has never been so high. When the carbon dioxide concentration reaches a level of 450 ppm, the temperature will rise by 2 degrees centigrade from pre-industrial period which is considered to be the maximum temperature we should not go beyond. Otherwise, extreme weather events and abrupt climate change will threaten the existence of human civilization on earth.

In 2015 at the COP21 climate change conference in Paris, the parties that met agreed that humans have raised the temperature and that this is not good. Recently climate scientist James Hansen published a paper showing that this has gone up 1 degree centigrade already. It will be miserable if we allow the temperature to rise 1.5 degree above that of the pre-industrial era. So we are working with a very small margin. If you take a look at the IPCC report during the pre-industrial era, everything is fairly stable and extreme weather events are not so frequent. Compare this to now; the possibility of extreme weather events is already very high. What this actually means is that the situation is really quite serious already. Extreme weather events are becoming more common and increasingly severe. So while we keep on saying that the upper limit is a two degree centigrade increase. This is quite irresponsible for future generations because a two-degree centigrade rise would put the earth system in a very dangerous situation. However if we continue the business as usual, it is possible the rise in temperature will be as much as 4.5 degrees at the end of this century, and the humanity is not likely to survive under these conditions. Unless we start a deep decarbonization immediately, and become carbon neutral by the middle of the century. The temperature rise is most likely to exceed 1.5 degrees.

As I said we are already in a very undesirable situation. Consider the Yolanda typhoon, also called Haiyan typhoon which landed on the Philippines in 2013 with a wind speed of 100 km/h or about 10 m/s, the most intense land falling tropical cyclone on record, killing 7000 people in one sweep. This type of weather related devastation is something which will happen with increasing frequency. When the leaders of 195 countries came to the COP21 meeting, they agreed that there was a need to swiftly decarbonize the energy system and negotiated a non-binding agreement for every country to work towards meeting a 2 degree centigrade upward limit. However it was said that ideally the limit should be set to a 1.5-degrees centigrade rise. Scientists will tell you that a 2-degree centigrade increase would be miserable. We really have to keep ourselves to this 1.5-degree centigrade as the upper limit. Other goals that were implemented included a pledge of one hundred billion dollars a year from developed countries towards helping developing countries, starting in 2020. This should be the relatively easy part of the agreement as global economic activity is around a hundred trillion dollars. This one hundred billion is a tiny sum of money. It was also agreed that in 2020 the carbon-reduction plans will be revisited.

What I'm saying here to you is that we are currently in a very dreadful predicament and even worse is that we do not have enough commitment to do things to save the next generation. I have served on the International Council of Science for many years as the President, this is an organization that is supposed to promote international science for the benefit of society. However, we are aware that in spite of the fact that civilization is heading towards a disaster, we are not doing enough. My advice to the younger generation is not to trust the older generation completely, when they tell you what to do, as they will not be there when the true crisis comes. The decision making should not be left to them.

Now I wish to summarize. Politically speaking, humanity is in the midst a global problem in need of global solutions. The United Nations however is nation-state based. There is no official mechanism to work together to implement guidelines for nations to do something really meaningful for humanity. Furthermore, from the social perspective, with too many people consuming too much and thereby overloading the earth, we are by definition over-developed. This over-developed world that has perilously taxed the planet does not have the right to dictate to the underdeveloped world the best way to proceed. The over-developed world needs to find different ways to reexamine so called development as well as finds means to turn down the impact of this. Today technologically the transformation from sunshine to electricity has become relatively inexpensive, but we have not yet learned how to store this electricity to make it available for the global market. So these are technological problems that we must learn to solve, transforming energy from the sun and storing this to make it available to the global market. Unfortunately, it seems that we are not progressing enough in this direction. We do not have a bright future ahead of us, and to change this we have to work hard, very hard. The first thing we have to do is to significantly decarbonize our society. By 2050, we must achieve carbon neutrality. While many scientists are doing everything they can, we need a total plan to realize this. Implementation of such a plan will not depend on science and technology alone. Society overall will have to transform. We must do this otherwise we will be very sorry.

If we want to decarbonize human society, we will need to depend on sunshine. The sun provides enormous amounts of energy. In 1 hour, the amount of sun transmitted energy to the surface of the earth is about equivalent to the energy we use in 1 year. While there are other energy sources such as methane hydrate which produces lots of energy, we are better off not to burn it. People are concerned that when the ocean warms that this might cause methane hydrate to start to bubble out. In the year 2000, a book was written that said that humanity will perish in 80 years and that part of the cause would be the increase of methane in the air as it released from the ocean and from the frozen tundra in Siberia. So it is not a good thing to use the methane hydrate. Instead we have to learn to store or transform the electricity generated by sunlight.

If you look at the current nuclear technology, it was built up more than 50 years ago and at that time and as it is today, people were worried about a proliferation of nuclear weapons. Current nuclear technology, from my point of view, is not a very ideal energy source. First of all, it is not safe enough. Secondly, we burn only about 2% of the uranium fuel and then 97–98% becomes waste, which is very difficult to manage. If every country were to work together, and come up with the next generation of nuclear power plant that would be safe and burn all the fuels and not produce radioactive waste, then you would not have the problem of weapons proliferation. This goal is not impossible.

In seeking out next generation technologies for sustainable environments science must remain aware of the threats that such technologies can also pose. You often hear mention of both the light as well as the shadows of science and technology. While science and technology caused human society to develop we also found out that with so many people pursuing material comfort, that we are changing the environment. I have participated in many international programs on science and technology, giving people the promise that science and technology will make this place better. However, looking at the entire world situation today we might say that uncontrolled applications of science and technology have led towards the negative, knowing that we have a critical planetary boundary restriction to satisfy. In the application of science and technology one has to be careful.

In response to those who question whether temperature changes are due to cyclical changes, if you trace temperature fluctuations backwards in time, you do see a glacial period, about every 20,000 years. This is because the orbital motion of the earth around the sun is not circular and the spin of the earth makes an angle to the axis of orbital motion. For a certain period there will be an ice period. But the time for this is in thousands of years. Some people will say if you wait, the glacial cooling period will come. But how long can we afford to wait? We are going to wipe out humanity within a hundred years if we allow business as usual, so how can we wait for thousands of years. Scientists will tell you that the sea level rise is irreversible now and by the end of the century, there will be a multi-meter rise. Whether its 1 m or 2 m depends on how effectively we are mitigating greenhouse emissions, but it can no longer be reversed. Scientists have studied every detail and although there are some more details that still need to be investigated, there is no longer a question as to the imbalance of energy coming in and going out and that the temperature of the surface of the earth is rising. This is why leaders of 195 Countries came together for COP21 because they realized the importance of this. The main problem is that circumstances and the ability to act continue to deteriorate.

Molecular Inference of Wisdom*

G. Terry Sharrer[1]

Wisdom is a human trait upon which natural selection creates evolutionary pathways from the molecular biology of the gene, to the cellular organization of nervous system, to the physical functioning of the brain — and the transit is a two way street. Wellness is a selection advantage at every level. Molecular wellness is innate, though often compromised because selection is blind; intellectual or high level wellness is learned behavior fed back to the molecular level as a kind of vision. The wisdom of this is that lifelong wellness — to the point of dying healthy — is a goal for every person and every society because it precedes and influences all other achievements.

Keywords: Wisdom; Wellness; Genomics; Well-Wise Neurons; Brain Enhancement; Evolution; Social Feedback.

*This chapter was based on a lecture delivered by Dr. Sharrer at the Molecular Frontiers Symposium and Youth Forum in the Royal Swedish Academy of Sciences in May 2014.

[1] Curator of Health Sciences (Ret.) Smithsonian Institution, Washington, DC, USA.

About Terry Sharrer

Terry Sharrer was curator of Health Sciences at the National Museum of American History, Smithsonian Institution, where over twenty years he built a renowned artifact collection in molecular medicine history, presented exhibitions at the Howard Hughes Medical Institute and Cold Spring Harbor Laboratory, published in medical history, and served on several medical foundation boards. Since 2007, he directed the Medical Innovation and Transformation Institute for a hospital group in Northern Virginia, and consulted with Cancer Treatment Centers of America. For a decade he has edited a weekly newsletter for MedicalAutomation.org and co-chaired several medical conferences. He has been a chair or speaker at other conferences at the Free University of Berlin, the Royal Swedish Academy of Sciences, Johns Hopkins University and George Mason University. His latest publication was a book chapter titled "Personalized Medicine: Ethical Aspects," in *Molecular Profiling: Methods and Procedures*, edited by Virginia Espina (New York: Humana Press, 2017).

Did you know that Fermilab, the world-renowned particle physics facility in Batavia, IL, outside of Chicago, has a bison herd on its campus? This seems curious for a place where work focuses on the discovery and measurement of subatomic particles. Their research has nothing to do with bison; rather the herd is there as a symbol of the frontier, just as it was for 25 years on the US nickel and still is today on the US$50 gold coin. Fermilab sees itself on the frontier of high energy physics. But this is not a simple coincidence. The lab's first director, Robert Wilson, was born in the town of Frontier in southwestern Wyoming and arose in his career during the flush times for science during the presidency of John Kennedy which was known as "The New Frontier." But, there's an earlier link. In 1893, at the Columbian Exposition in Chicago, historian Frederick Jackson Turner presented a talk that had profound and long lasting influence on American intellectual life. Its title was "The Significance of the Frontier in American History."

Turner argued that pioneers moving into an undiscovered space created a new expression of democracy, with both good and ill characteristic, such as a strong desire for independence, along with a high degree of violence. He portrayed the frontier experience as a kind of evolution. Later generations modified but did not wholly supplant the theme that a frontier, open for opportunity, is a desirable condition that becomes more complex over time. Today, we search for new frontiers and they exist among the stars, in the unexplored depths of Earth's oceans, and in ourselves. Self-knowledge may be the ultimate frontier. In this talk, I'll propose a frontier thesis of the brain as a way highlighting what earlier speakers on this program have "pioneered." First, though, let me explore with you the molecular inference of wisdom.

Wisdom is an observable human trait and always has been, or at least since Confucius, a half millennium before Christ. Confucius wrote "By three methods we may learn wisdom: first, by reflection, which is the noblest; second, by imitation, which is the easiest; and third, by experience, which is the bitterest." He also noted "Wisdom, compassion and courage are the three universally recognized moral qualities of men." By all accounts Confucius possessed and displayed great wisdom in his "day job" as an administrator of agricultural affairs. His recorded genealogy is the oldest of any person, coming to the present with a boy born in Taiwan in 2006. It would be interesting to know if wisdom has any measurable heritability through 80 generations.

Besides being observable, wisdom is highly distributed in nearly all people. Rene Descartes made a joke of this, pointing out that the most widely distributed trait in the world also means no one brain possess very much of it. While we know that wisdom is highly variable, it comes in two intellectual "flavors:" the individual, and the cultural. On the low end of individual wisdom is the folk expression, "common sense," and one of the great purveyors of this form was Benjamin Franklin in the utilitarian pamphlets he printed annually between 1732 and 1758, known as "Poor Richard's Almanack." Among dozens of homely homilies he compiled were "Early to bed, early to rise, makes a man healthy, wealthy and wise," and "Drink does not drown Care, but waters it, and makes it grow faster." He once summarized these as "the Thirteen Virtues," which, like "the Ten Commandments" offer wisdom for every day private life.

But individual wisdom also has a higher order in things like "inspiration," "intuition," and "insight." It may be possible, as the 19th century phrenologists supposed, that some forms of thinking are localized in the brain, but that seem improbable. Higher orders likely involve the entire brain and for now remain as mysterious to us as the oldest light in the universe, though the latest finding about the oldest light was just announced last month. It comes from the flash within the first trillionth of a trillionth of a trillionth of a second after the "Big Bang," and it has been traveling outward for about 13.8 billion years. An expanding rather than contracting universe is supposedly a good thing.

Scientific inquiry, of course, has been a great source of wisdom. Galileo's observation about our sun-centered solar system produced the fundamental wisdom that, as much as some wanted to believe that the sun revolves around Earth, science proved otherwise. But science isn't infallible, as we can see in a 1927 US Supreme Court case, "Buck vs. Bell."

Carrie Buck was a young woman, whose mother had a criminal record for prostitution. Both lived in Augusta County, VA. When Carrie was known to the local authorities to have had a daughter out of wedlock, she was charged with "feeblemindedness" under

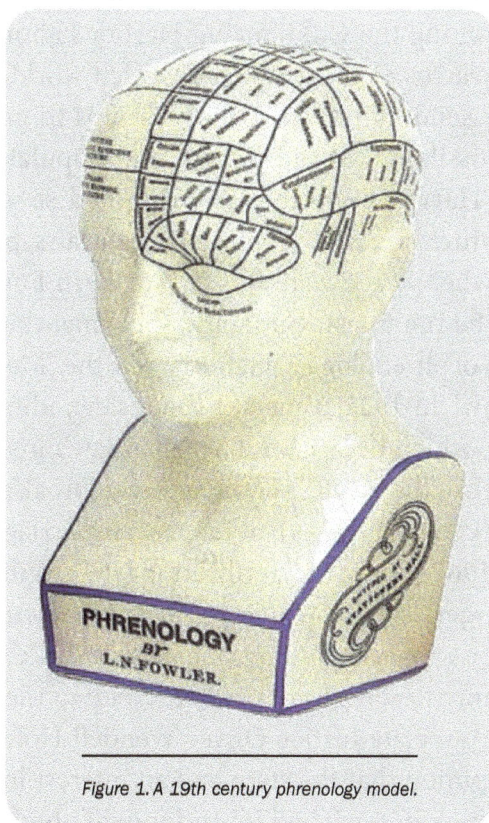

Figure 1. A 19th century phrenology model.

the 1924 Virginia eugenics law that prescribed compulsory sterilization for people deemed intellectually disabled. This law was the product of some of the most scientific advanced "wisdom" of its day. The term eugenics originated with Charles Darwin's cousin, the biometrician Sir Francis Galton in 1883. It meant "well born." Of course, it also implied that some were not so well born, and questions arose over how to detect and measure fitness and what to do about the "unfit." Some of the early thinking about solutions were castration and spaying, just as farmers did with livestock. An American, zoologist Charles Davenport from the University of Chicago became a leading proponent of eugenics research and policy. In 1910, he established the Eugenics Record Office on the grounds of the biological research station at Cold Spring Harbor, NY (later becoming the Cold Spring Harbor Laboratory — one of the major hearths of the genetics research world). Among his activities were sending out field study groups to gather data on curious and possibly unfit individuals and populations. He and his successor Harry Laughlin also promoted so-called "Fitter Families for Future Firesides" contests at state fairs, particularly in the Midwest, where there were lots of Northern European stock, considered to be the fittest. Generally speaking, this was a more advanced form of phrenology. Laughlin wrote the "Model Eugenical Sterilization law" in 1922, aimed at the insane, idiots, imbeciles, and epileptics, which became the Virginia law's inspiration. Not stopping with that, the Virginia General Assembly also legislated its "Racial Integrity Act" which forbad interracial marriages until it was overturned in 1967. Under the sterilization law, Virginia performed "neutering" surgeries on 6,683 people (4,043 women; 2,640 men). Most were poor whites, including one, Carrie Buck. Her case had gone from county to state courts, and then up to the US Supreme Court, where Associate Justice Oliver Wendell Holmes, Jr. wrote the majority opinion that the state had an interest in the society's "gene pool" which outweighed an individual's body privacy. He thought compulsory sterilization was no different from compulsory vaccination. And, supposing that Carrie, her mother and daughter were probably all feebleminded, he concluded with the memorable line "three generations of imbeciles are enough."

The Eugenics Movement was not confined to Virginia or even the United States. Many countries including Canada, Japan, Belgium, Brazil and Sweden had laws that resulted in sterilization of people with psychiatric disorders. Eugenics became the foundation of Nazi Germany's race hygiene policies and the ultimate destination for this line of scientific thinking was the gates of Auschwitz. Because race hygiene became a rationale for genocide, eugenics gradually faded away. Virginia repealed its eugenics sterilization law in 1974; Sweden did so the following year. "Ethnic cleansing" is the little brother of racial hygiene where limited murder rather than full blown genocide aims to drive certain populations out of a society, such as happened with the alleged cannibalism against Mbuti Pygmies during the Congo Civil War in 2003, and the

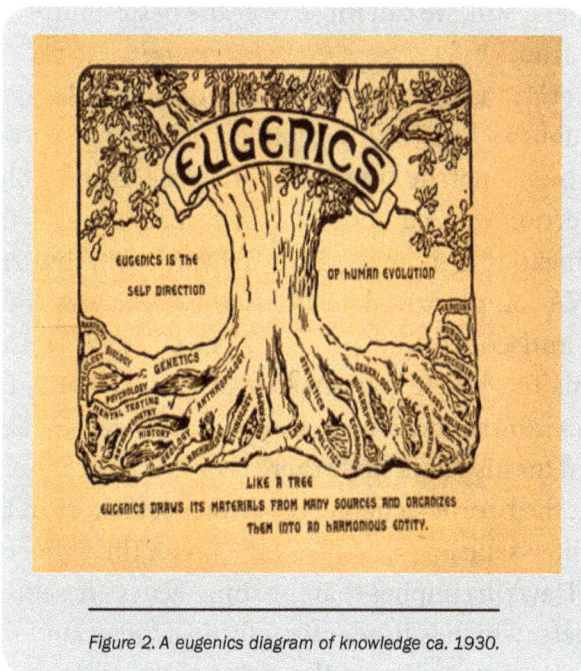

Figure 2. A eugenics diagram of knowledge ca. 1930.

South Sudan conflict between the Dinka and Nuer peoples which is going on today. Only superficially are these eugenic conflicts, but they follow in the wake of ideas that science once considered "wise."

Individual and cultural wisdom interact with each other, and while they seem to be at least qualitatively different, they have a common origin in our genomes. So, let's begin looking for molecular inference there even though it involves enormous complexity (quite the opposite of the genetic reductionist approach). The number of genes in the human genome is thought to be around 21,000, down considerably since the last decade when most scientists believed the number was around 100,000. Maybe there are fewer yet — perhaps an "essential, unduplicated" set, but in any case it isn't a finite number of genes that accounts for a genome, but rather an epigenetic array of hyper-astronomical combinations and modifications, in the realm of 10 followed by millions of zeroes. It's arguable which is greater, epigenomic biochemistry or the number of neural connections, but it doesn't matter; both are unimaginably large numbers. Still, we can muse over the basic nature of the gene and the neuron.

A generation ago, British evolutionary biologist Richard Dawkins publish a small and deceptively plain-language book titled *The Selfish Gene*. In it, he argued that the gene, as the molecule that natural selection worked upon, was selfish about its own survival. He didn't mean this in an anthropomorphic way, but rather that units of DNA competed and the fittest for a particular environment prevailed and continued on to the next generation. Classic Darwinism. The less fit perished or at least remained dormant until the environment changed. Dawkins has been challenged and even vilified for his views since there is philosophical weight to the notion that the bedrock character of life is selfish, even though he explained how selfishness can lead to autism through complexity. In a sense, Dawkins implies that the thing genes are selfish about is their wellness, which is fundamentally an adaptation for survival. Ironically, this comes close to the well-born meaning of eugenics.

Figure 3. The human genome repository at the Sanger Center, Cambridge University, UK. Image courtesy of the Sanger Center.

About a decade after *The Selfish Gene*, geneticist Christopher Wills at the University of California in San Diego published a book proposing that "the rapid evolution of our intellects has been facilitated by the ways in which the genes controlling development of our brains are organized. And, as with everything else in evolution, increases in intellectual ability have happened more than once." The title he chose was *The Wisdom of the Genes*. Like Dawkins he didn't imply an anthropromorphic "wisdom," but rather that the accumulated ability of genes to evolve in some cases and not evolve in others, represents a fundamental wisdom. This, of course, happens in the context of cells — neurons, for our purposes — and it's possible that the neurons of the autonomic nervous system are better at not evolving because they are always "on" for maintaining blood pressure, respiration, and other things, while the neurons of the frontal lobes are better at evolving for intellectual performance. Wise genes evolve into well genomes, which include well-wise neurons. I'm using wellness and wisdom as interchangeable terms, because that's the molecular inference.

In 1958, Francis Crick gave us what's called "the central dogma of molecular biology" which, simply put, is DNA makes RNA makes protein. We now know that this pathway is a two way street, and almost daily neuroscientists report congested traffic patterns. Earlier this spring, for example, the FANTOM consortium, headquartered in Japan reported how expression of the same gene differs according to its tissue location. What varies is the point in messenger RNA where protein synthesis begins. The gene for the dopamine active transporter protein, for example, is located on human chromosome 5, with 15 exons, spanning some 64,000 base pairs. Messenger RNA transcribes this gene, but the translation into protein can begin at different points depending on whether the expression is taking place in, say, the cerebral cortex, cerebellum or elsewhere. Mistranslations, polymorphisms and post translational modifications, such as methylation, are possible causes of or contributors to bipolar disorder, depression, and attention deficit-hyperactivity syndrome. These conditions may inhibit wisdom, prevent it, direct it into a different pathway, or conceivably even promote it, as might have been the case with depressives like Abraham Lincoln, Charles Dickens and Leo Tolstoy, or biopolar sufferers such as Ludwig von Beethoven, Winston Churchill and Isaac Newton.

One more point about traffic patterns. Pyramidal neurons are cells found in the brain's cerebral cortex, hippocampus and amygdala, which are implicated in the cellular basis of wisdom. Santiago Ramon y Cajal, the Spanish pathologist called "the father of neuroscience" and winner of the Nobel Prize in 1906, discovered pyramidal cells and illustrated them as having a triangular shaped axon, with a long dendrite extending from its apex that branches as it moves out, and shorter base dendrites with branches. Both dendrites have spines which increase the impulse input from other cells while the axon serves output. Over the past several years, a group at Ludwig-Maximilians University in Munich has figured out how that input works. Microtubes run the length of the dendrites and axon, and function much as highways. Motor proteins called kinesin move molecular cargoes in one direction, toward the axon tip; while another, dynein, move them in the opposite direction. The cargoes are neuronal ribonucleoprotein granules which release their messenger RNA component at specific dendrite spines. These messenger molecules result in protein synthesis the reconfigure the shape of synapses. As the particles transit from one dendrite to another the sites of messenger RNA release become repeatedly activated in developing a memory for the delivery route. Even here, driving and texting don't happen simultaneously without dangerous consequences. In fact, the driving proteins repress the "texting" RNA from activating, which indicates that transport and translation are discrete processes. Otherwise, RNA-binding proteins might crash into what we call Alzheimer's and Parkinson's diseases. Over time, even without accidents, the delivery vehicles, their cargoes, and even the highways show wear and tear, which are part of natural aging. Environmental factors can alter this traffic pattern, but like highway safety, the wisdom of the genome has wellness as its original intent.

Researchers at Ludwig-Maximilians University who gave us insights into the ribonucleoprotein traffic patterns relied on antibodies to sort out the proteins of those granules. This technique built upon the work Gerald Edelman did at Rockefeller University in the 1960's and for which he and Oxford University biochemist Rodney Porter shared the 1972 Nobel Prize. In proving that antibodies are made of variable heavy and light chains of amino acids, Edelman and Porter explained how the immune system works. Further, with each cell division slight changes in the DNA coding generated further variation which natural selection worked upon to achieve in Edelman's words, not survival of the fittest, but survival on average of the fitter — that is, a more gradual rate of evolution. He might well have spent the rest of his career in immunology, but the model of antibodies binding to cell surface antigens took him down another path to the discovery of adhesion molecules on cell surfaces which allow cells to form tissue. As subsequent studies showed there are four major families of cell adhesion molecules, the largest being the 765 members of the Immunoglobulin superfamily, which besides adhering cells to one another carry out signaling for growth and development including synapse formation. Here is another level of diversity for natural selection to act upon, and Edelman thought that neural evolution worked through three kinds of selection with one constraint. He saw "developmental selection" creating functional but heterogeneous neuronal groups, suggesting that no two people had exactly the same synapse construction; then, "experiential selection" from which memory develops involving multiple locations in the brain; and "reentry" where development and experience interact, resulting in a higher order of selection. Just as antibodies can react to antigens they have not encountered before, higher order selection produces the brain capability that transcends the molecular to the cellular level. He also added one more factor — value system which constrain selection processes by introducing biases, one of which may be the Dawkins' "selfish" impulse. An important implication of Edelman's interpretation of neural Darwinism is the molecular inference that wisdom is widely distributed within the brain — virtually a whole brain phenomenon, which isn't surprising except that since the 19th century localization has been a main theme of neuroscience.

We are so sure that the brain is the seat of all mental processes that it's hard to imagine that for most of history that wasn't the case. Aristotle, contradicting both his teacher Plato and Hippocrates, proposed that the heart guided emotions and possibly the heart, liver, spleen and kidneys were responsible for all things biological. He thought the brain was simply a radiator that cooled the heart. Aristotle's interpretation became what was known as "humoral theory" which held that a careful balance of four fluids — blood, phlegm, black bile and yellow bile — explained wellness and imbalance, illness. From this model comes the heart representing love, and sayings like "keep a cool head," and "gut intuition." Today, when we muse over melancholy as Robert Burton did in 1621 in the first book about the anatomy of melancholy, we are using a word that literally means black bile. Humoral theory gradually eroded from explorations of the nervous system, the brain's topography and the mind's "nature," but it was a philosophical map from the individual to the cosmos for about 2,500 years.

In the 19th century, along with the scientific revolutions that arose from Darwin, Mendel, and Pasteur, we have a theory called "phrenology." German physician Franz Joseph Gall was the first popularizer of the notion that bumps on the skull, which resulted from bone protrusions into the brain, revealed specific functions of the mind. He concluded that the brain actually was not one organ, but rather 27, each with a specific location and mode of activity. With a slightly later proponent, Johann Spruzheim, phrenology spread in Europe and America, and with it localization interpretations of the brain and the mind's moral concepts. This has an interesting history that leads into eugenics, but that's a story for another day. Fast forwarding to the present, we have two California psychiatrists publishing in *The Archives of General Psychiatry* (2009) on the "Neurobiology of Wisdom?: A Literature Overview." Actually, with fluorescent markers like GCamp6, and functional neuroimaging, the literature traffic is getting heavy. PubMed has more than 26,000 full text journal articles under "wisdom," though more than a few are by authors named Wisdom — Jennifer, Ronald, Amy and William, who often are neuroscientists of one kind or another. After their literature overview, Thomas Meeks and Dilip Jeste reached these conclusions which I'll quote:

> "The prefrontal cortex figures prominently in several wisdom
> subcomponents (e.g. emotional regulation, decision-making,
> value relativism, primarily via top-down regulation of limbic

and striatal regions. The lateral prefrontal cortex facilitates calculated, reason-based decision-making, whereas the medial prefrontal cortex is implicated in emotional valance and prosocial attitudes/behaviors. Reward neurocircuitry (ventral striatum, nucleus accumbens) also appears important for promoting prosocial attitudes/behaviors. Monoaminergic activity (especially dopaminergic and serotonergic), influenced by several genetic polymorphisms, is critical to certain subcomponents of wisdom such as emotional regulation (including impulse control), decision-making, and prosocial behaviors.... Wisdom may involve optimal balance between functions of phylogenetically more primitive brain regions (limbic system) and newer ones (prefrontal cortex). Limitations of the putative model are stressed."

It would be unfair to simply characterize these observations as old phrenology way-finding on the new sophisticated map of human brain science. But it does illustrate that from antiquity to our time philosophers and scientists have argued over whether or not the brain and the mind were one in the same or separate, whether mental processes are localized or distributed over the whole brain,

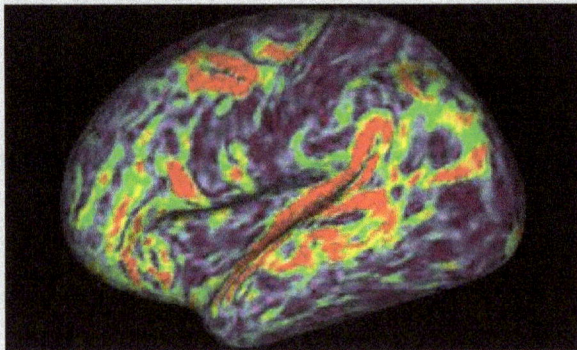

Figure 4. A PET scan showing 97 regions of the brain recently discovered. Image courtesy of The Berkshire Eagle (http://www.berkshireeagle.com/stories/updated-brain-map-identifies-nearly-100-new-regions,201373).

and whether or not neurons alone explain consciousness and concepts like wisdom. These debates have lasted so long because the ultimate question is whether or not the concept of the soul is real. If the molecular inference of wisdom is selfish wellness, souls don't need an explanation. Molecular actions explain everything. If it's altruism, inference supports religious belief, which is a direction Meeks and Jeste move in whether intentionally or not. Still, to give religion its due, it was Job's "Poem of Wisdom" in the Old Testament (Job, 27:12) which asked "But where shall wisdom be found? And where is the place of understanding?" Job, Ecclesiastes and Proverbs make up the Old Testament's so-called "wisdom literature." It's worth knowing these three books don't entirely agree on the means to wisdom.

Between Job's question and our understanding of neurobiology, the context of wisdom and wellness which interests us most is the individual, specifically personality, because it's closest to our desire for self-knowledge. Personality is what distinguished people from things. It's not a single quality, but an assemblage and often seen in negative terms. In the Oscar Pistorius murder trial, for instance, the prosecutor painted Pistorius as narcissistic, without empathy, paranoid and a liar — trying to establish quilt by personality.

It's safe to say that personality emerges from brain chemistry, in which neurotransmitters are the pace-setters. German-born pharmacologist Otto Loewi discovered the first neurotransmitter in 1921, but interest grew after World War II, in part due to genocide committed with nerve gas. Two neurotransmitters that seem crucial to presenting wisdom are serotonin and dopamine — the former first explained in the work of Betty Twarog at the Cleveland Clinic in 1953, and the latter, by Arvid Carlsson at Lund University in 1957. Before discussing these, we must again recognize the complexity involved. Multiple genes are involved in dopamine and serotonin expression, emitting, reception and control — as they are encoded on two chromosomes in each instance. Mutations, promoters, copy number variations and post translational modifications of the DNA induce variation from the start. Then there are generalizations about their action that are often vague or contradictory. So, we have to explore with caution.

Fifteen years of exploration followed Twarog's discovery which turned out to be a pathway to socialization. In the brain, serotonin influences mood, and plays a role in memory and learning. Too much of it has been linked to obsessive-compulsive disorder; too little, to depression. The antidepressant drugs Prozac and Zoloft work by retarding serotonin reabsorption in neurons which diminishes depression to the point of reversing it to a sense of well-being. That is the simple version of a protein which is generated in cells of the brain stem. It does its work through other proteins: the serotonin transporter which transits the extracellular space to at least seven different cell surface receptors, and monoamine oxidase-A which breaks down serotonin. In the case of the transporter, a gene on Chromosome 17, has a promoter sequence of 22 DNA bases, with either fourteen or sixteen repeats — the so-called "short," low expressing, and "long" high expressing versions. Studies have shown that the sixteen repeats result in what the ancients would have called a "sanguine" personality, coming from more serotonin circulating in the brain. The monamine oxidase-A gene, on the X chromosome, has thirty bases in its promoter region which are repeated between two and six times. High expression levels have been linked to depression disorders, low expressions (resulting in greater amounts of serotonin, dopamine and noradrenaline) have been associated with impulsive aggression and violence.

Dopamine and serotonin are both monoamines — dopamine deriving from tyrosine; serotonin from tryptophan. Both are degraded by monamine oxidase-A. And, as with serotonin, too much or too little dopamine has behavioral consequences. A deficiency can "freeze" personality; short of that, it causes indecisiveness. The best known deficiency is the absence of dopamine in the substantia nigra which causes Parkinson's disease. People with heightened dopamine levels are thought to be adventurous, thrill-seekers, and schizophrenic at hyper levels. The dopamine receptor gene, on Chromosome 11, has copy number repeats, where the greater the number the less effective the receptor is at capturing dopamine. Too much dopamine may explain what drives parachutists to jump off skyscrapers or people in barrels to go over Niagara Falls.

Autism, in some cases, derives from too much dopamine. Autistic symptoms can appear during infancy, when the baby won't hold eye contact with its mother or does not want to be held. A little older and they remain emotionless, except for fits of rage when their compulsive routines are disrupted. As adolescents they do not seek kindness from others, nor do they show compassion to others. They usually have difficulties with language — in learning to read, word use, and speech. "Autism" went unnamed until 1938, when Hans Asperger in Vienna coined the term, meaning self, though Martin Luther may have encountered an autistic child whom he thought was Devil-possessed, and the Wild Boy of Aveyron fits the symptomatic description. This was a lost child of the French Revolution who lived alone for seven years in the wilderness. He was about twelve when of his own accord he came into a town. A young physician, Jean Itard, took him in, and named him Victor. Curiously, Victor became a kind of celebrity of the European Enlightenment, as natural philosophers speculated over qualities that distinguished people from animals. Itard specifically looked for language ability and empathy in Victor; he found some evidence, but not much of either. Carl Linnaeus speculated that Victor might belong to another species, "Homo ferus" (a feral human). Rousseau and Decartes dug for a deeper "nature" of humans, as in Decartes's famous line "I think, therefore I am." But in Victor's case, the philosophers who found wisdom in nature did not take him as a reasoning human who could be civilized. He was a kind of sub-human, though above the animals. On a larger scale, the same interpretation supported empire-building with so-called "uncivilized" native peoples' lives. Victor died in Paris at age 42.

It seems quite probable that Victor was autistic, but impossible to know why. Dopamine is implicated with some autistics, but so too is serotonin. Researchers at Kuopio University in Finland found autistic children who presented low amounts of the serotonin transporter. The term "autism spectrum disorder" includes both, but also implicates other neurotransmitters, monoamine oxidase-A, cortisol resulting from stress, gonadal hormones, and perhaps hundreds of other proteins in the brain's traffic patterns. But, in addition, there are social factors.

Psychiatrists have observed that people who achieve a self-satisfying status — through their work or social achievements — have correspondingly high levels of serotonin. This has been found in animal studies too. A reversal in status, perhaps diminishing self-esteem, lowers serotonin. Similarly, nurturing autistic children can help ameliorate behavior by lowering their dopamine levels. A few years before Arvid Carlsson identified dopamine as a neurotransmitter, Johns Hopkins University psychiatrist Leo Kanner proposed that autism was rooted in a "genuine lack of maternal warmth." This became known as the "refrigerator mother" theory which other psychiatrists fanned into thinking that autism was a parenting disorder. In some cases, it may have been. An abusive parent who humiliated, intimidated, or otherwise injured a child's self-esteem — especially a child who was born somewhere on the natural spectrum of autism — would have switched genes off or on, driving to disorder. "Refrigerator" mothers and fathers could have been autistics themselves. By the 1980's, however, research about neurotransmitters and the organizing of parents with autistic children into advocacy groups and foundations rebalanced the interpretation to nature and nurture. Nature provides the genes; nurture affects their expression levels.

Earlier I mentioned Thomas Meeks and Dilip's literature review, "the neurobiology of wisdom," in which they defined wisdom as a composite of six components: (1) prosocial attitudes/behaviors; (2) social decision-making/pragmatic knowledge of life; (3) emotional homeostasis; (4) reflection/self-understanding; (5) value relativism/tolerance; and (6) acknowledgement of and dealing effectively with uncertainty. If this holds up we can define autism as the thief of wisdom. Severely autistic children show none of these traits; in fact, they show polar opposites.

You'll also recall that Richard Dawkins proposed "the selfish gene," where natural selection gave a survival advantage to selfish genes that were the best adapted to their environment. Christopher Wills wrote about "the wisdom of the genes," which is close to Dawkins' argument, their common ground being wellness abetting survival. Still, how could selfish genes end up supporting the kind of altruism (selflessness) Meeks and Dilip define as wisdom?

On the last page of his book, Dawkins backed off a little from a single, unremitting selfishness by speculating about a possible difference between selfish genes' short term interests and long term ones. A short term gain could inhibit a long term one, such as might arise between an illness and aging. Aging is not simply accumulated wear and tear on cells; illness and aging involve two distinct processes, though they are linked by a finite number of cell divisions. But in either case, cells function in a distributed, but highly connected network. Dawkins characterized this as a "conspiracy of doves," where genes give up a bit of their innate short term selfishness for a long term advantage in wellness.

If, as I propose, wisdom and wellness are synonyms, let's briefly consider what wellness has come to mean. Wellness definitions stretch from "an active process of becoming aware of and making choices toward a more successful existence" to a physio-psycho balancing act akin to homeostasis, but all who have addressed the subject agree that wellness is not the opposite of illness. Wellness is a higher state of health. In the simplest sense, it is a combination of health and happiness — neither of which are easier to define than wellness itself. The preamble of the World Health Organization's Constitution (1946) reads that "health is a state of complete physical, mental and social well-being and not merely the absence of disease or infirmity." The meaning of "well-being," however, remains vague, even while it has observable physical and psychological dimensions. Social wellness, for example, emphasizes human interactions, noting that isolated individuals are more susceptible to depression and weakened immune responses, leading to a greater premature death rate. Medical wellness recognizes that many actors play upon health and, consequently, there are multiple complementary and alternative approaches to achieving a higher well-being. What all dimensions — social, physical, intellectual, emotional, mental, medical — have in common is self-motivation — an active pursuit where, in effect, every person strives toward self-reliance, even while depending on commercial products and services from others.

The former chief innovation officer for the US Centers for Disease Control and Prevention once pointed out the relative weight of factors contributing to premature death: genetics 30%, social 15%, environmental 5%, medical care 10%, and behavioral patterns 40%. All of these involve some degree of variability, but none more so than the choices we make in behavior. All people learn how to make good choices, which largely represent survival advantages. One of the hallmarks of developed countries is that the range of choices and the number of people who can make choices constantly expands — what Swedish historians Folke and Karin Dovring called "the optional society." It may be that this widening range of choice contributes to the incidence of so many people making bad choices in their behavior, but perhaps that's a fundamental question that remains on the molecular frontier.

Summing up the molecular inference of wisdom comes to this. Wisdom is a human trait upon which natural selection creates evolutionary pathways from the molecular biology of the gene, to the cellular organization of nervous system, to the physical functioning of the brain — and the transit is a two way street. Wellness is a selection advantage at every level. Molecular wellness is innate, though often compromised because selection is blind;

intellectual or high level wellness is learned behavior fed back to the molecular level as a kind of vision. The wisdom of this is that lifelong wellness — to the point of dying healthy — is a goal for every person and every society because it precedes and influences all other achievements. This does not imply we should attempt to vastly expand human life expectancy. That would be quite unwise as it gives pathogens a standing target to attack; extinction becomes the price of immortality. But within our limited number of cell division spanning a maximum of about 120 years, advancing human wellness would be wise enough.

A frontier thesis of the brain could explore wisdom, but way finding poses three questions: what are the pathways to follow; what's needed to carry out the exploration; and how will discovery change the pioneers and those who come afterward?

The pathways are really about what might be discovered, and in the complexity of the brain the possibilities approach the infinite, much like exploring the universe. I'll only mention three:

— what is the pathway to personalized, predictive medicine? This arises from the "omics" revolution — genomics, proteomics, metabolomics, *et al.* — which shows the uniqueness of each person (i.e. you are unique just like everyone else) and what is your "fate." A gene expression study of medullablastomas in children illustrates this point.

— what is the pathway to redefining diseases in finer detail than what is obvious? Hippocrates described epilepsy and for three thousand years it has been understood as storms in the brain. Henry Molaison, patient "H.M." to neuroscientists since 1953, was a 23 year old epileptic who the anterior two thirds of his hippocampus surgically removed to control his seizures. In a more radical procedure, a three-year old Maryland boy had a hemispherectomy for the same reason. The seizures were obvious; their cause was not, even leaving open the question of whether or not epilepsy is one disease or many under one name.

The same might be applied to many, even all, brain and mind disorders — which might also extend to obesity, hypertension, and all kinds of addiction.

— what is the pathway to a more benevolent society? Every society is becoming more diverse which can strain the concept that civility depends on a disposition to do good. The "golden rule" — do unto others as you would have them do unto you — is the essence of benevolence. But diversity also introduces unfamiliar attitudes and values that often are seen as threatening, as we find in racism and ethnic discrimination. Could an innate xenophobia be the basis for this, and if so, how can it be managed? If not, what causes persecutions, from bullying to crime to mass murder? Eugenics attempted to answer this; so, we should expect this pathway to be a rough road.

Neuroscience will need new means to find its way on the frontier. Again, there are many and earlier speakers on this program have discussed one of most exciting discovery tools, optogenetics. I'll mention three others: gene therapy, neural implants, and astute family doctors.

In March 1993, Kevin Klug from Cincinnati received gene therapy for a glioblastoma multiforma — one of the most serious forms of all cancers. He had surgery, chemotherapy and radiation therapy before, but the tumor grew back each time. Pioneering gene therapist at the US National Institutes of Health, Michael Blaese, drew up the protocol which relied on a recombinant retrovirus that incorporated the Thymidine kinase gene from Herpes simplex into the engineered Moloney Murine Leukemia Virus. After surgeons cleared the vascular bed, Dr. Blaese introduced the virus, which could only infect dividing cells as would have remained from the tumor, unseen. As you know, Herpes simplex causes cold sores, and anti-Herpes work by acting on the Thymidine kinase gene. So, in effect, gene therapy allowed this kind of brain cancer to be converted to a cold sore which was then treated with the anti-viral,

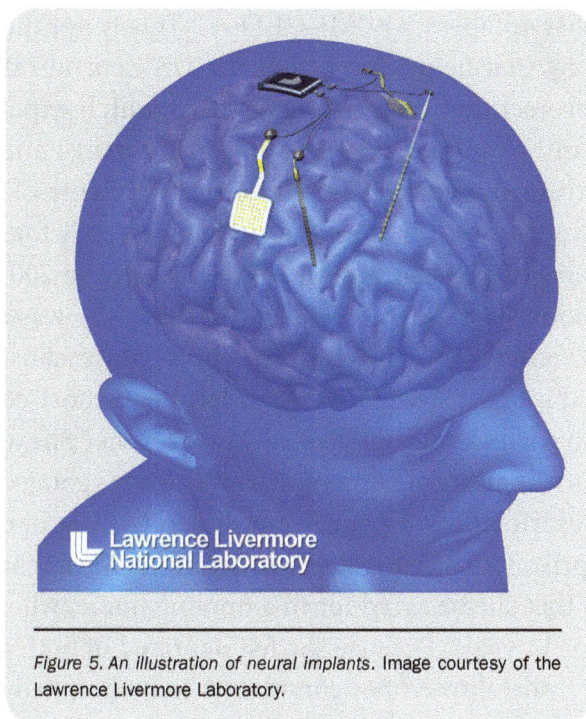

Figure 5. An illustration of neural implants. Image courtesy of the Lawrence Livermore Laboratory.

Thymidine kinase-targeting drug Gancyclovir. Kevin, who had a 30 day life expectancy when this was done, is now out 21 years. Gene therapy is a complicated technology that requires finding the best viral vectors, the proper producer cells to expand the virus, and careful delivery. Among the neurotropic vectors used today are HIV and rabies, and certainly gene therapy is on the frontier.

When George Orwell published "Nineteen Eighty Four" in 1949, neural implants were conceivable but impossible. In fact, they had been imaginable since H. G. Wells published "A Modern Utopia" in 1905, and conjured further in Aldous Huxley's "Brave New World" in 1931. All three of these works of fiction figure in the social background of neural implants because they represent more than a century's fear of such things. Their social acceptability is preconditioned by notions of psychological manipulation, mind control and brainwashing, all of which have non-fiction realities, perhaps most notably with "MKULTRA." This is not the Chicago rock band by that name, but rather the US Central Intelligence Agency's project, between 1953 and 1973 which experimented with human behavioral engineering, using drugs and various kinds of torture. It involved 149 subprojects which the CIA secretly contracted out to universities and private research foundations. Files that survived a burn order were declassified in 2001.

Now comes Cathy Hutchinson. In 1997, she was a 42 year old single mother of two, living in Attleboro, Massachusetts when she suffered a brain stem stroke that left her almost completely paralyzed in what's known as "locked-in syndrome." She was unable to move, but was aware of events and conversations going on around her. She described losing her brain, but retaining her mind as being in a prison for a crime she didn't commit. Massachusetts law prescribed that she be cared for in a nursing home, where in fact, she lived for 11 years. Because she felt deprived of both good care and privacy, she allowed her complaints to be bundled in a class action lawsuit, Hutchinson v. Patrick, in 2007 which prevailed and has since allowed nearly 1,200 institutionalized brain injured people to move out of nursing homes and into community residences where they can be closer to friends and family.

But that's not the end of Ms. Hutchinson's story. After winning the lawsuit, Ms. Hutchinson became a pioneering patient for neural implants. Dr. John Donoghue of Brown University and colleagues at Massachusetts General Hospital created a device called "BrainGate." It is a tiny 100-electrode array of nanowires and silicon tips that interface with neurons of the motor cortex on the top of the brain. The implanted sensor detects the neurons' electrical transmissions, which normally control motions, and sends them to an external decoder where software translates thoughts into the mechanical movements of a robotic arm. A remarkable YouTube video for BrainGate shows Ms. Hutchinson thinking her way through drinking from a thermos of coffee for the first time in 15 years. Donoghue is now working on a wireless version for this technology.

BrainGate, along with cochlear and retinal implants and devices for deep brain stimulation to control Parkinson's disease are enabling technologies for the impaired. Neural implants might also be developed for memory loss, drug addiction, or bipolar disorder. From there it's not hard to imagine cyborg soldiers or athletes with performance-enhancing brain chips. Proceeding further to neural implants having a status akin to cosmetic surgery for those who could afford it, and government regulation that sees this technology as an acceptable competitive advantage, we could, indeed, have a "brave new world" for better or worse.

It is reasonable speculation that gene sequencing, optogenetics and neuron mapping, neural implants and "silent talk" brain imaging will allow neuroscientists to discover new pathways that capture headline news. But these technical wonders will wither on history's vines if they do not reach the practices of astute primary care physicians who can scale big science into small science. Small sciences is where doctor and patient meet face to face in search of illness individuality. An example of such a place is the Clinic for Special Children in Strasburg, PA. It is the primary care facility for the descendants of 16th century German-speaking Bohemian and Swiss Anabaptists who to migrated to America and became known by the folk term "Pennsylvania Dutch." They are the Amish and Old Oder Mennonites. Because of the so-called "founder effect" and faith-based marriage traditions, today's populations have high incidences of inherited metabolic errors that are rare in the general population. Maple Syrup Urine Disease, for example, is found in about 1 in 120,000 live births worldwide, but 1 in 271 live births among the Groffdale Conference of Old Order Mennonites in Lancaster County, PA. This disease floods the basal ganglia with the three branched chain amino acids, leucine, isoleucine and valine, and, in effect, starves the mitochondria of other essential amino acids, leading to basal ganglia death, if left untreated. Treatment involves an infant formula without the branch chain amino acid, and careful monitoring, especially during periods of typical childhood illnesses. (These slides show 71 such illnesses, of which 40 appear in Amish families, but not Mennonite, and 31 among Mennonites, but not Amish.) Being able to detect these diseases before irreparable damage is done depends on having highly sophisticated genomic and proteomic equipment in the Clinic's laboratory and having three physicians and a geneticist on staff who are familiar with population-specific molecular medicine and can practice it every day. Not many clinics have patients arriving by horse and buggy, but humanity has many genetic bottlenecks, and their relatively simple, if devastating, inherited illnesses are components, to some extent, in every other human disease. Like wisdom, between gene and cortex, big and small science represent a two way feedback mechanism.

Lastly in this frontier thesis of neurobiology, we have to consider how pioneering in the brain has changed us already and might do so further in the future.

Obviously, discovery has led to new drugs for brain diseases. Further research could produce medicines that are personalized to an individual's metabolic rate, more targeted and with fewer side effects or adverse reactions. Beyond that, though, neuro-discovery could, and almost certainly will, redefine the nature of illness and wellness. In genetic and/or cellular pathways, it may turn out that some Parkinson's sufferers have more in common with some autistics and some schizophrenics than they do with other Parkinson's patients. Already, neurologists speak about "synaptopathies," as for example, "your form of Parkinson's is a disfunction of glutamate transmission rather than dopamine, and is actually a rarer form of Fragile X syndrome." Getting the diagnosis right in the first place is a big step in making health a more achievable and sustainable goal. Since it's easy to see that most of what we call modern times would have been impossible without the discovery of germs, we can muse about previously unimaginable possibilities stemming from molecular brain theory.

In his book, "World Wide Mind," science writer Michael Chorost raises a plausible, if not yet possible, interface between the human brain and the internet. Part of his inspiration comes from having two microcomputers in the form of cochlear implants connected to his auditory nerves; another part comes from his addiction, like millions of other people, to electronic "smart" devices that provide frequent stimulation, frequent distraction, and unconscious slipping away from direct personal contact — that is "real touch" as opposed to "touchscreen." With self-assembling microrobots that could form wireless electrodes at specific functional zones on the brain's cortex, it might become possible to put on a "thinking cap" as a wearable internet. This could avoid both real touch and touchscreen, resulting is an even greater isolation than we now see with a young couple, on a date, each paying more attention to their smart phones than to each other. Before that becomes a social pathology, however, people will have invented new forms of enchantment and intimacy, as Chorost says, that are necessary for survival of both individuals and societies. These new forms, which certainly should bring new values and behaviors, might be troubling or be fun, but either way they will be irreversible.

I'll now turn the frontier thesis of the brain over to you. Think about the possibilities of brain function enhancement that could improve communications between people, or improve communications between the individual's conscious and subconscious self. When your generation has mapped the brain, start mapping the mind. And when you reach these intersections, remember the last stanza of American poet Robert Frost's "The Road Not Taken:"

Two roads diverged in a wood, and I —
I took the one less traveled by,
And that has made all the difference.

The Origin of Life
on Earth and the Design
of Alternative Life Forms*

Jack W. Szostak[1]

To understand the origin of life on Earth, and to evaluate the potential for life on exoplanets, we must understand the pathways that lead from chemistry to biology. Recent experiments suggest that a chemically rich environment that provides the building blocks of membranes, nucleic acids and peptides, along with sources of chemical energy, could result in the emergence of replicating, evolving cells. The broad scope of synthetic chemistry suggests that it may be possible to design and construct artificial life forms based upon a very different biochemistry than that of existing biology.

Keywords: Origin of Life; Prebiotic Chemistry; Protocell; RNA Replication.

---

* This chapter was based on a lecture delivered by Professor Szostak at the Molecular Frontiers Symposium in the Royal Swedish Academy of Sciences in May 2017.
[1] Nobel Laureate in Physiology or Medicine (2009). Investigator, Howard Hughes Medical Institute; Professor of Genetics, Harvard Medical School; Professor of Chemistry and Chemical Biology, Harvard University; Alex. A. Rich Distinguished Investigator, Department of Molecular Biology, Massachusetts General Hospital.

About Jack W. Szostak

Jack W. Szostak (born November 9, 1952, London, England) is an American biochemist and geneticist who was awarded the 2009 Nobel Prize for Physiology or Medicine, along with American molecular biologists Elizabeth H. Blackburn and Carol W. Greider, for his discoveries concerning the function of telomeres (of DNA occurring at the ends of chromosomes), which play a vital role in determining cell life span. Szostak also investigated the process of chromosomal recombination during cell division and conducted studies into the role of RNA in the evolution of life on early Earth. In addition to Szostak's investigations into telomeres, he was the first to create a yeast artificial chromosome (1983), which can be used to clone DNA and consists of a vector (or carrier) molecule that contains yeast genes necessary for replication and a DNA segment of interest.

Szostak received his B.Sc. from McGill University in Montreal in 1972, and then conducted his graduate research under the supervision of Ray Wu at Cornell University, Ithaca, NY, obtaining his Ph.D. in 1977. Szostak then moved to the Sidney Farber Cancer Institute and Harvard Medical School in 1979, and then to Massachusetts General Hospital in 1984.

In the 1990s, Szostak and his colleagues developed in vitro selection as a tool for the isolation of functional RNA, DNA and protein molecules from large pools of random sequences. His laboratory used in vitro selection and directed evolution to isolate and characterize numerous nucleic acid sequences with specific ligand binding and catalytic properties. Using only simple molecules, he developed techniques to generate functional RNAs in a test tube. The goal of this research was to synthesize a self-replicating protocell susceptible to Darwinian evolution, which could then serve as a model to investigate the transition from chemical to biological life on early Earth. For this work, Szostak was awarded, along with Gerald Joyce, the 1994 National Academy of Sciences Award in Molecular Biology and the 1997 Sigrist Prize from the University of Bern. In 2000, Szostak was awarded the Medal of the Genetics Society of America, and in 2008 Szostak received the H.P. Heineken Prize in Biophysics and Biochemistry.

From 2000 until the present, Szostak's research interests have focused on the laboratory synthesis of self-replicating systems and the origin of life. For this work he received the Harold Urey Medal from the International Society for the Study of the Origin of Life in 2011.

Szostak is an Investigator of the Howard Hughes Medical Institute, Professor of Genetics at Harvard Medical School, Professor of Chemistry and Chemical Biology at Harvard University, and the Alex Rich Distinguished Investigator in the Department of Molecular Biology and the Center for Computational and Integrative Biology at Massachusetts General Hospital.

Szostak is a member of the National Academy of Sciences and the American Philosophical Society, and a Fellow of the New York Academy of Sciences, the American Academy of Arts and Sciences, and the American Association for the Advancement of Science.

One of the things that I particularly love about the subject of how life got started is that there are so many interesting questions, and that they cover such a wide range of fields, from astronomy to geology to chemistry and biology. Just getting to talk to people from so many different fields and trying to figure out how everything came together to create the conditions for the emergence of life is really fascinating. The approach to understanding the world by always asking new and interesting questions is one of the great things about science, and the origins field is particularly special in the way that it brings people together to cooperate and collaborate and try to get answers to some of the mysteries of nature. When we see so many divisive and tragic things happening in the world today it's really wonderful that we're privileged to be able to work together.

Let me begin with a brief account of how I came to be interested in the subject of the origin of life. In the 1990s my lab spent a lot of time learning how to evolve populations of molecules[1]. It was a fascinating endeavor, and we along with others including Gerald Joyce and Larry Gold, learned how to evolve new molecules that could do interesting things such as bind to targets and catalyze reactions. Some of these experiments even led to molecules that are now in the clinic to treat diseases. But after evolving molecules in the lab for roughly a decade I became more and more fascinated by the question of how evolution started spontaneously on the early Earth[2]. It's one thing to direct evolution in the lab when you have enzymes and expensive instruments and brilliant students. But somehow the process of Darwinian evolution got started all by itself on the early Earth. Since my original background is in biology, to me the origins of evolution and the origins of biology are really the same thing. As a result I started to focus more and more on the issues surrounding the origin of life, and that is the subject of this article.

What might a really simple primitive cell (or protocell) look like? One model is that protocells consisted of a primitive membrane enclosing some very short pieces of genetic material

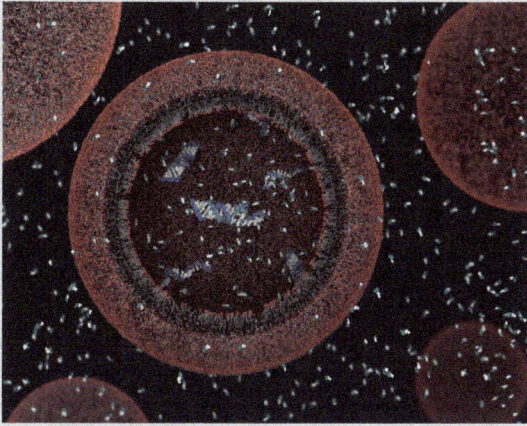

Figure 1. Illustration of a model protocell, consisting of a lipid membrane encapsulating small pieces of an RNA-like genetic material. The protocell is shown in a rich chemical environment filled with nucleotides and other building blocks of life. Image courtesy of Janet Iwasa.

(Figure 1). We're not totally sure of the nature of that first genetic material, but for many reasons we think it is probably closely related to RNA[3]. What we're trying to understand is how structures like this came to exist on the early Earth, and how they were able to grow and divide and start to evolve.

People have been asking how we got here, i.e. how life got started, for a long time, but one of the things that has generated an explosion of interest in this question is that we now know that there are millions, probably hundreds of millions of planets that are reasonably Earth like, in our galaxy alone. We are pretty sure that many of them could support life, but we don't know if they do. An example of a particularly interesting nearby solar system is the TRAPPIST-1 system that has been worked on a great deal by Didier Queloz and his colleagues[4] in Cambridge in the U.K. (Figure 2). This solar system has at least seven planets and three of them are in the zone (green in the figure) where there could be liquid water on the surface. It's quite likely that planets like this

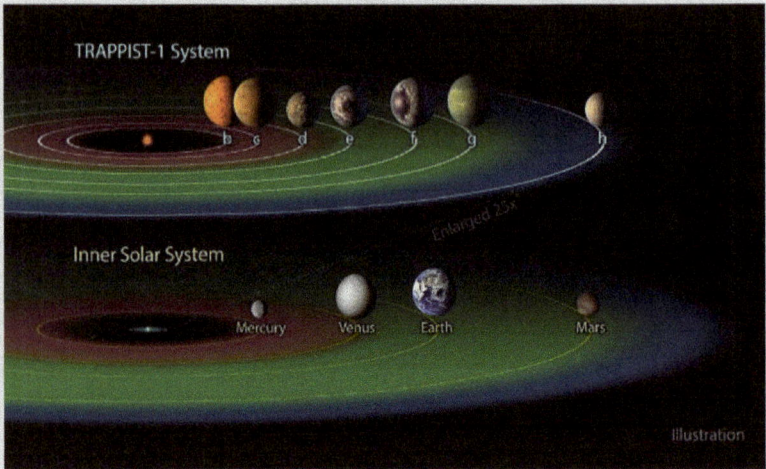

Figure 2. The TRAPPIST-1 planetary system is one of the many recently discovered planetary systems orbiting other stars in our galaxy. Three of the seven planets of this system lie within the habitable zone, where liquid water might exist on the planetary surface. It is likely that such planets could support life, at least in local environments. Detecting signs of life on exoplanets is one of the great questions energizing current astronomical research.

could support life, but it is going to be really hard to figure out if they are actually inhabited. The entire field of astronomy has been revolutionized by this kind of discovery and scientists are trying as hard as they can to work out the technology to detect signs of life on these exoplanets.

In parallel with that ongoing effort, we can ask simple questions that can be addressed here in our laboratories on earth, to try to understand how life might get started on different kinds of exoplanets, and how it did start on our planet. If we understood the whole process we could make a better estimate of how easy or hard it is for life to begin, and therefore how likely or unlikely it is that there is life out there. This involves a lot of questions on the whole pathway from planet formation to the early evolution of life, but here I will focus on how life could actually get started from the chemistry on a young planet.

We first need to understand how the building blocks of biology might be synthesized on a young planet. The study of this question is the field of prebiotic chemistry, and it is being pursued intensively in a number of laboratories. Then, once we have the right chemicals, we need to understand how a set of molecules can get together and start acting like a living cell. That is the topic that we have been focusing on in my laboratory. Recently, work in these two areas has advanced to the point that we can start to deduce something about the necessary environments for the origin of life[5]. This is bringing us into contact with planetary scientists, geologists, and atmospheric chemists to try to be as rigorous as we can in thinking about early planetary environments.

Let us begin by considering some of the hard problems with prebiotic chemistry. Stanley Miller's famous apparatus, built in 1952, used a spark discharge as a source of energy in what was then thought to be a realistic primitive atmosphere[6]. In this simple experiment, Miller made an astonishing number of different compounds, but most surprisingly amino acids, one of the quintessential types of building blocks of biology. This was a revolutionary advance at the time, and Miller's experiment generated huge optimism that the chemistry leading to life would soon be figured out. There was in fact a lot of progress in the ensuing 10 or 20 years. And then progress stalled and some of the problems with this approach started to become more evident[7]. In particular if you start to look at what is in this mixture of compounds, the closer you look the more you see. There are thousands, probably tens of thousands of different chemicals made in this crude approach of blasting a reducing atmosphere with energy. A lot of the materials we want are in there but at very low levels, while some of the materials we know we need are not there.

How can we think in a more realistic way about generating the building blocks that we need to put cells together? In my opinion a series of new approaches to this problem have transformed the field. I will try to give an impression of how the field has changed in the last 10 to 20 years as a result of these new approaches. The idea that one simple process could generate a primordial soup that contained everything needed for life is clearly unrealistic; what we need to figure out is how to make complex chemical pathways possible. Since it takes multiple steps to get to something as complicated as the nucleotides needed to make RNA, it is not realistic to expect this to happen by just mixing simple starting materials together. But you can break a pathway into smaller chunks if you can build up reservoirs of intermediates as crystalline deposits — in effect, organic minerals. Another approach is systems chemistry[8], which goes beyond the classical chemical approach of taking two compounds and making them react to produce a product that you want, by thinking about what else could or must be present and that could make a difficult step work in a better way.

Here I will give a few examples of potential organic minerals, and of systems chemistry approaches to the problem of nucleotide synthesis. One of the ideas that has been around for a long time is that life began from cyanide. Cyanide is not good for us now but it is a fantastic starting material for the chemical synthesis of all the building blocks of life[9]. But how would cyanide be made on the young earth? It is known that many different sources of energy (UV, lightning, impacts) acting on a primitive atmosphere can generate cyanide. The trouble is, how do you actually make use of cyanide that forms in the atmosphere? If it rains out into the ocean it will just hydrolyze and be lost. However John Sutherland has devised a scenario[10] for how to capture, store and accumulate cyanide, and then later turn it into useful products (Figure 3). The idea is that cyanide rains out onto the surface and collects in lakes or ponds in an area where there is geothermal circulation of the water through fractured rocks. Water circulating through rocks and heated by the magma deep below comes up through vents and brings up ferrous iron and other ions. Ferrous iron reacts very rapidly with cyanide to make ferrocyanide. Certain ferrocyanide salts are insoluble and will precipitate. One can imagine building up layer after layer of

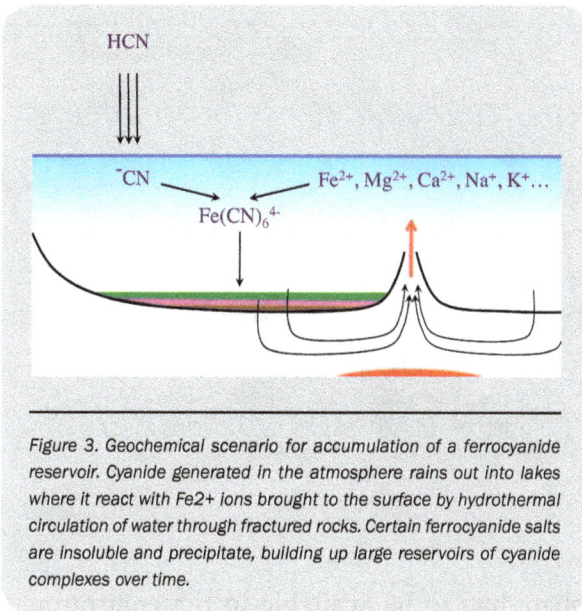

Figure 3. Geochemical scenario for accumulation of a ferrocyanide reservoir. Cyanide generated in the atmosphere rains out into lakes where it react with Fe2+ ions brought to the surface by hydrothermal circulation of water through fractured rocks. Certain ferrocyanide salts are insoluble and precipitate, building up large reservoirs of cyanide complexes over time.

insoluble ferrocyanide salts over thousands of years. As a result, very dilute cyanide could potentially be converted into a huge reservoir of material that could subsequently be processed by heat and water to make useful starting materials.

Another example, also from the Sutherland lab, illustrates the principles of systems chemistry. The compound 2-aminooxazole (2AO) doesn't look particularly biological but it's very important because it is a potential prebiotic intermediate on the way to building nucleotides and RNA[11]. It turns out that 2AO can be made from two very simple starting materials: the simplest sugar, glycolaldehyde, and the cyanide derivative cyanamide (Figure 4). If you just mix them together, not much interesting happens because you get a huge range of products and polymeric materials, and just a little bit of the compound that we'd like to have. What the Sutherland lab found is that if you put phosphate in the reaction mixture, it works much better. Now phosphate is not one of the starting materials, and it's not in the product, but we know

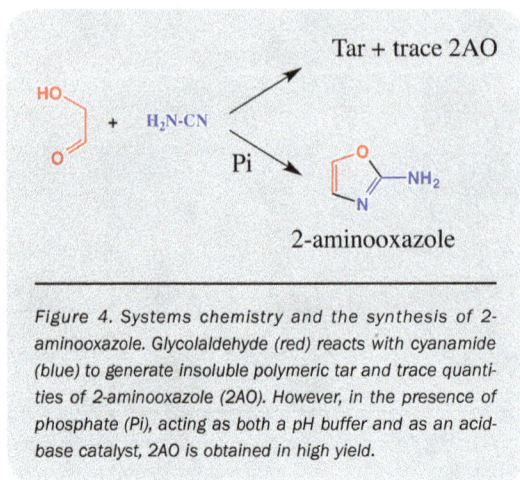

Figure 4. Systems chemistry and the synthesis of 2-aminooxazole. Glycolaldehyde (red) reacts with cyanamide (blue) to generate insoluble polymeric tar and trace quantities of 2-aminooxazole (2AO). However, in the presence of phosphate (Pi), acting as both a pH buffer and as an acid-base catalyst, 2AO is obtained in high yield.

that phosphate had to be available in the environment because phosphate is part of nucleotides and it is part of RNA. Phosphate is both a good pH buffer and a good acid-base catalyst, and it turns out that in the presence of phosphate our two starting materials react rapidly to make much more of the key intermediate 2AO. This is an excellent example of systems chemistry: using molecules that aren't directly part of the reaction, but that should be present and that can make the chemistry work better. Remarkably, it turns out that 2AO is volatile, so it sublimes at moderate temperatures and then crystallizes on nearby cold surfaces. As a result it could self-purify and accumulate as a crystalline reservoir.

Another simple physical way of purifying an intermediate is by crystallization from the reaction mixture. In the very next step on the pathway to nucleotide synthesis, 2AO reacts with the sugar glyceraldehyde to form the key intermediate ribose aminooxazoline (RAO). Remarkably, RAO crystallizes out of solution (Figure 5), leaving behind left over starting materials as well as all of the other products are made[12,13]. You can imagine this occurring on a geological scale, and building up a large reservoir of this crystalline intermediate. This is a perfectly natural and plausible way of

Figure 5. Crystallization of ribose aminooxazoline (RAO). 2AO reacts with glyceraldehyde to form RAO, a precursor of the pyrimidine nucleotides, but the arabinose, lyxose and xylose aminooxazolines are also formed. Only RAO crystallizes from the reaction mixture. Adapted from Hein, Tse and Blackmond, Nat. Chem., 2011.

building up a reservoir of an intermediate so that if conditions change in the right way, the next step in the pathway can happen.

The examples given above highlight some of the new ways of looking at prebiotic chemistry that make it easier to understand how complicated compounds like nucleotides could actually be formed on the early Earth. But what happens then? Suppose we have all the amino acids, peptides, nucleotides and lipids we need: we still have to understand how these molecules get together to make a living cell. We think a very simple stripped down version of a cell would have looked like this: essentially a primitive cell membrane enclosing little fragments of something like RNA (Figure 6)[14]. Here the nature of the questions that we're asking is different. How could such a cell assemble, and then manage growth and division? Since we are talking about the first cells, there can't have been any evolved biological machinery to coordinate these processes. Therefore there had to be very simple chemical and physical processes that

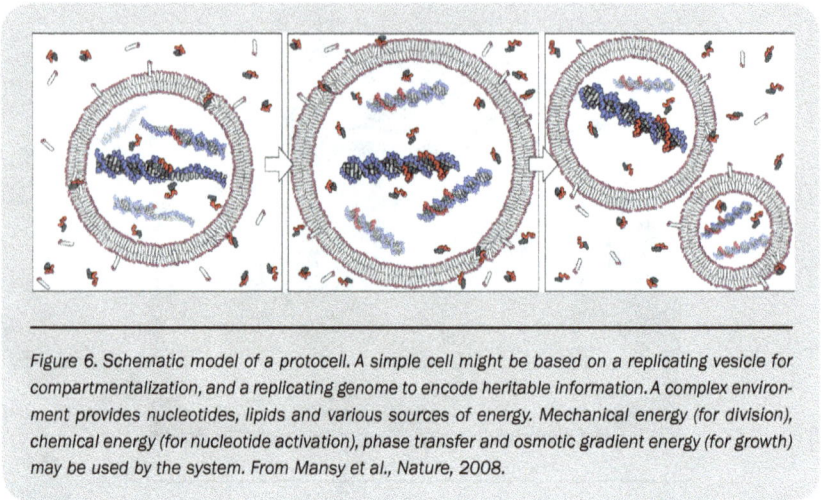

Figure 6. Schematic model of a protocell. A simple cell might be based on a replicating vesicle for compartmentalization, and a replicating genome to encode heritable information. A complex environment provides nucleotides, lipids and various sources of energy. Mechanical energy (for division), chemical energy (for nucleotide activation), phase transfer and osmotic gradient energy (for growth) may be used by the system. From Mansy et al., Nature, 2008.

could drive the primordial cell cycle. The key questions are: how did the primitive cell membrane grow and divide, and how did the genetic material replicate without enzymes? These are the questions that we have been working on for the last 10–20 years; here I will discuss just a few highlights from our efforts. The membranes of primitive cells were certainly not as complicated as the modern cell membranes surrounding our cells; primitive cell membranes had to have been made of simpler materials, and must have had very different properties. One possibility that we and others[15] have explored is that primitive membranes might have been made from fatty acids, which easily self-assemble into membranes. For example, if you start with oleic acid, the fatty acid derived from olive oil and commonly found in soap, and shake it up in water, the fatty acid molecules will spontaneously make bilayer membranes that close up into beautiful spherical vesicles (Figure 7).

The spontaneous assembly of fatty acid vesicles is reasonably well understood, but what is truly remarkable but far less well understood is that fatty acid vesicles can grow and divide as a result of simple physical processes. What you can see in some of our videos are vesicles with a fatty acid membrane and a fluorescent dye on the

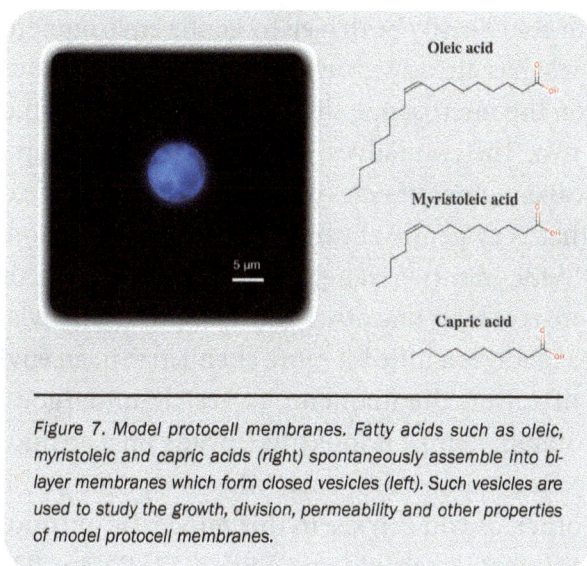

Figure 7. Model protocell membranes. Fatty acids such as oleic, myristoleic and capric acids (right) spontaneously assemble into bilayer membranes which form closed vesicles (left). Such vesicles are used to study the growth, division, permeability and other properties of model protocell membranes.

inside. We can watch in the microscope and see what happens when we add more food, i.e. more fatty acids[16]. We expected the initial vesicle, which we think of as a model of a protocell membrane, to gradually get bigger, perhaps somewhat elongated if the surface grew faster than the volume. What we didn't expect was what actually happened. If you watch the movie (http://molbio.mgh.harvard.edu/szostakweb/videos/Zhu_2009_JACS_01.qt) you will see a thin filament growing out of the originally spherical vesicle, and this filament grows, getting longer and thicker while the contents of the original vesicle gradually spread throughout the filament. This process looks very biological because there is a lot of movement and the shape of the filament changes as it grows. But amazingly this is just soap and water, nothing else. One of the beautiful things about this mode of growth is that the filaments are very fragile, which makes division very easy. At the beginning of our work on modeling protocell division, we had no idea how to make vesicles divide in a way that could easily happen on the early earth. However, once we saw how vesicles actually grow, it turned out that division could easily be driven by gentle environmental forces. Once a vesicle has already grown into a long filamentous shape, a puff of air on the microscope slide is enough to cause the filament to snap in two. The two halves will gradually round up and form two spherical daughter vesicles over time. That kind of experiment taught us that very gentle shear forces are sufficient to cause one vesicle to divide into two daughter vesicles after growth. It turns out that there is at least one other way that protocell division could work[17]. This involves a little bit more chemistry: in an environment with a lot of sulfur, the filamentous vesicle undergoes a strong pearling instability, until it looks like beads on a string, and then the little beads separate from each other over time and float away from each other, as you can see in this movie: (http://molbio.mgh. harvard.edu/szostakweb/videos/Zhu_2012_PNAS_02.mov). In this remarkable process many small daughter cells are generated from one initial cell. Satisfyingly, by combining these different approaches to growth and division, it is possible to make a vesicle grow into a filament and then divide into multiple daughters, then grow and divide again, and so on, indefinitely. In terms of a primitive cell membrane, we therefore have an environmentally controlled cycle of growth and division. In fact we now have many ways of driving growth in addition to the two different ways of driving division. All of this makes the cell-cycle like processes of growth and division appear to be quite simple in that they can be driven by very gentle and common physical forces.

Next, we need to consider the nature of the primordial genetic material. In order for primitive cells to be able to evolve and thus become better adapted to their environment, they had to have some way of having new functions arise, and those new functions had to be coded in a genetic material so that they could be passed on from generation to generation. The idea that RNA molecules played this early role is quite popular, in part because RNA does so many things in modern biology. One of the most important of those modern functions is protein synthesis: it is actually the RNA components of the ribosome that are responsible for synthesizing all of the proteins in our bodies, and indeed in all living cells on the modern earth[18]. This is a strong rationale for believing that ribozymes existed before protein enzymes. As a result, we think that protocells used RNA to catalyze important reactions such as RNA replication. But before there were ribozymes that could make RNA replicate more efficiently, there must have been chemistry that could have driven the copying of RNA sequences. How might that have worked? In order to illustrate the process, Janet Iwasa prepared a very nice animation to illustrate the nonenzymatic copying of an RNA template, which you can watch here: (http://exploringorigins. org/downloads/nonEnzymaticReplication_narr.mov). The idea is that there were short strands of RNA floating around in a rich chemical environment filled with activated building blocks. These activated nucleotides find their partners by base-pairing, G with C, and A with U, and then click together, building up a complementary strand. This process of chemical copying is easy to imagine and to illustrate, but demonstrating this experimentally is still an unsolved problem. People have been thinking about and working on this problem for a long time.

Much of the most important early work on chemical copying of RNA was done by the late Leslie Orgel and his students and colleagues including Jerry Joyce[7]. I think Orgel's greatest single contribution was to recognize that the nucleoside triphosphates that are used for DNA replication and for making RNA in all modern life are great for cells that have sophisticated enzymes to catalyze polymerization, but without such enzymes, our modern substrates can't do much except sit there and slowly hydrolyze. So, at the beginning of life, nucleoside triphosphates are not appropriate building blocks. Leslie therefore searched for and found different kinds of activation chemistry that work much better and allow for template copying without enzymes (Figure 8). In the best of those molecules that were first made in the Orgel lab[19], two of the phosphates of the triphosphate moiety of modern NTPs have

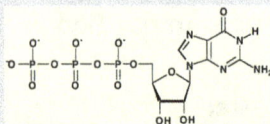

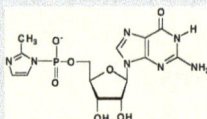

Figure 8. Modern vs. prebiotic means of nucleotide activation. Modern cells use nucleoside triphosphates (NTPs, top) for the enzymatic synthesis of RNA and DNA. Primitive cells must have used more reactive and less polar substrates, such as the nucleoside phosphorimidazolides (bottom), for RNA replication.

been replaced with a 2-methylimidazole group. As a result, these nucleotides are much more reactive. They don't need an enzyme to polymerize, so they can copy simple RNA sequences without an enzyme.

The reactive nucleotides described above can be used to copy short stretches of RNA. In a typical experiment we begin with a primer, for example a little piece of RNA six nucleotides long, that is base paired to a template, which is a somewhat longer piece of RNA. The goal of the experiment is to copy a string of Cs in the template strand by extending the 6-mer primer using activated G building blocks. The Gs should pair with Cs, and then they will be in position to react with the primer, as a result of which the primer should grow longer. In a time course we start off with just the primer, the six nucleotide RNA. After a minute, the first G nucleotide has been added. A few minutes later the second one, then a few minutes after that, the third G has been added to the primer. This is amazing chemistry: there is no enzyme but the RNA template is being copied! Well, that's the good news about this chemistry. The bad news is that is the only reaction that works really well. If we try to copy a sequence that has all four nucleotides in it, it basically doesn't work. We can put in all four activated building blocks and let them sit with the primer and template and nothing happens.

Why is copying mixed sequence templates so hard? Are we thinking about this in the wrong way, and is there some missing chemistry? Due to lack of space I cannot review all of the interesting advances that have been made in tackling this problem[20], so I will skip ahead to discuss one advance from last year in my lab. Remarkably, after 30 years of thinking about and working on this problem, we discovered that if we change the methyl group on the imidazole to an amino group, and make the new leaving group 2-aminoimidazole (2AI) (Figure 9), the copying chemistry works much better[21]. As you can see, the pace of progress in this field is sometimes a little slow, but nonetheless, progress is being made. These new 2AI activated monomers are nice because they speed up the copying chemistry; now when we try to extend an RNA primer by copying a template, we can completely copy up to seven nucleotides and sometimes a bit more. We still need to make this chemistry work even better, but we have many ideas about how to copy longer sequences. If we could copy templates that were 20 to 30 nucleotides long, that might be

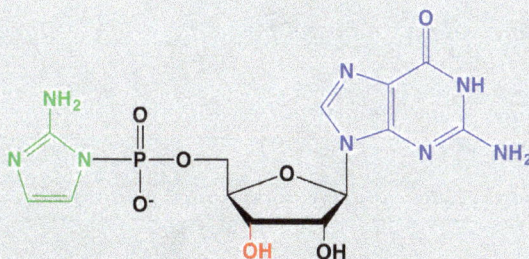

Figure 9. A new leaving group for activated nucleotides: 2-aminoimidazole (2AO). Nucleotides activated with 2-aminoimidazole lead to faster and more efficient primer extension than nucleotides activated with 2-methylimidazole.

good enough to bootstrap evolutionary processes within synthetic protocells, because we know that we can make good catalytic RNAs by assembling complexes from chunks of RNA that are 20 to 30 nucleotides long. That is our current goal.

Let's return to our new activating group, 2-aminoimidazole (2AI). One question that arises is whether this is just something that works in the lab, or does it have a plausible connection with the origin of life? Would molecules like that have been made in the early chemistry of our planet? This time we think that it might really be relevant. The reason we think so is shown in Figure 10. On the left is 2-aminooxazole, the intermediate that goes on to build nucleotides, as worked out in the Sutherland lab[11]. On the right is 2-aminoimidazole (2AI), which we use to activate nucleotides so that they will polymerize. These two molecules differ only in the identity of one atom — where there is an oxygen in 2AO there is a nitrogen in 2AI. It turns that 2 AI can be made in the same reaction mixture as 2AO just by adding ammonia[22]. Even more amazingly, there is another very similar molecule, 2-aminothiazole (2AT), in which the oxygen of 2AO is replaced with sulfur. This compound can react

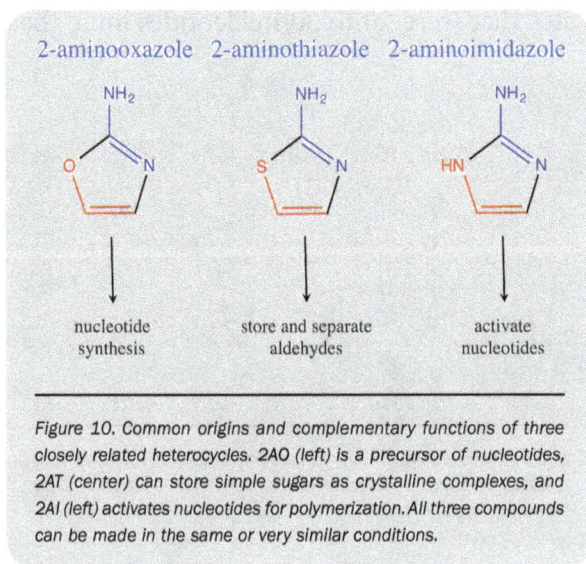

Figure 10. Common origins and complementary functions of three closely related heterocycles. 2AO (left) is a precursor of nucleotides, 2AT (center) can store simple sugars as crystalline complexes, and 2AI (left) activates nucleotides for polymerization. All three compounds can be made in the same or very similar conditions.

with and store, in a stable crystalline form, the simple 2 and 3 carbon sugars that are needed to make nucleotides and amino acids[23]. I find it amazing that these three very closely related molecules, all made by simple chemistry from simple starting materials in the same or very similar environments, can go on to play different roles in building RNA molecules — this seems like a strong clue that this chemistry is truly relevant to the prebiotic chemistry of the early earth.

What did the primitive genetic material look like? Despite the issues discussed above, I still think that the first genetic material was very similar to modern RNA. Many polymers similar to but distinct from RNA have been made in the laboratory and studied, but so far nothing that looks like a better candidate than RNA has emerged. Similarly, recent advances in prebiotic chemistry begin to suggest how RNA might have arisen, but so far don't point to alternatives. As a result we think the primordial genetic material was more or less the same as RNA, but probably a messier version with some heterogeneity (Figure 11). For example, the nucleotides

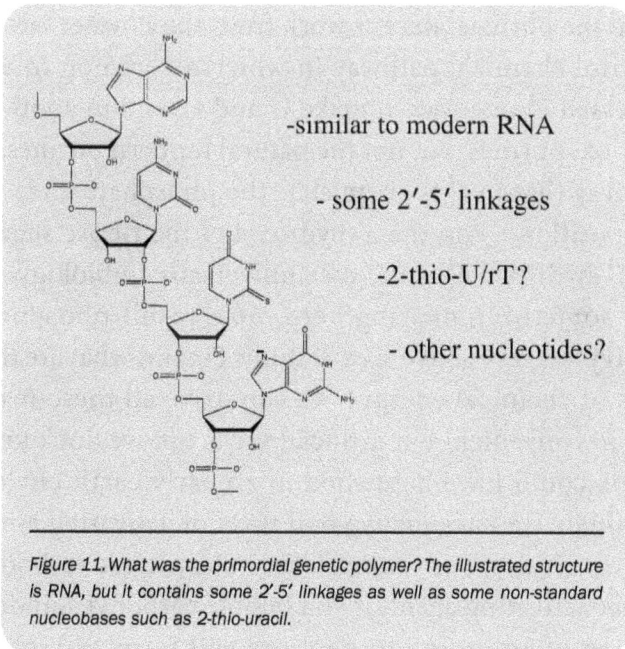

-similar to modern RNA

- some 2′-5′ linkages

-2-thio-U/rT?

- other nucleotides?

Figure 11. What was the primordial genetic polymer? The illustrated structure is RNA, but it contains some 2′-5′ linkages as well as some non-standard nucleobases such as 2-thio-uracil.

in primitive RNA were probably not always linked up in the 'correct' way. We know that different backbone linkages (i.e. 2′-5′ instead of the biologically universal 3′-5′) can form during chemical copying, but we have recently found that this variation is not as harmful as we used to think[24]. It is also possible that there could have been small chemical changes in some of the nucleotides. For example, replacing O_2 of uracil with sulfur makes the RNA copying chemistry work much better, and this change is actually still seen in biology in part of the protein synthesis apparatus[25]. There could also be other small changes, yet to be discovered. But overall, the structure of primitive RNA is likely to have been close enough to modern RNA that it could easily evolve step by step into what we see now in biology.

What is still missing from all this chemistry? There has been a huge amount of progress in prebiotic chemistry, but there are still major gaps in our understanding. From the Sutherland lab we think we know how to make the pyrimidine nucleotides U and C in an efficient and plausible manner[26], but we still don't know how to make the purines. Recent work from the Powner lab describes a beautiful chemical pathway in which a common intermediate is processed in one way to make U and C, and in another way to make 8-oxopurines, but not the natural modern purines. Another problem is that in this chemistry, the phosphate ends up in the wrong position — on the 3′-hydroxyl of the ribose sugar instead of the 5′-hydroxyl as is the case universally in biology, so there's clearly something missing here. As a result phosphorylation chemistry is a very active area of study. Finally, what are the correct sources of chemical energy? We can drive all these reactions in the lab in convenient but artificial ways, but we don't understand how this could have happened in an early earth environment. Interestingly we have many good ways of activating amino acids to make peptides. For example, the volcanic gas carbonyl sulfide is a great activating agent. If you bubble carbonyl sulfide through a solution of amino acids, peptides will form and lots of other interesting chemistry will happen as well[27,28]. So, the search is on for an analogous way to activate nucleotides so they can engage in template copying chemistry and polymerize without enzymes.

What environments on the early earth could have supported the chemistry that gave rise to biology? And what environments could have nurtured the first very primitive cells, and driven their growth, division and evolution? We think that life began in geological settings where organic compounds could accumulate to high concentrations. If useful molecules are made, but fall into the ocean, they will be diluted and lost. This makes surface lakes or ponds seem like more interesting possibilities to consider. Figure 12 is a picture of Yellowstone Lake in the western US. Perhaps in a similar environment on the early earth, organic materials could have built up over long periods of time, generating a richly concentrated chemical environment. We also think that the environment where

Figure 12. Yellowstone Lake in winter. Organic compounds could accumulate in similar lakes and ponds on the early earth, and proximity to the atmosphere along with exposure to UV light could have supplied sources of chemical energy. Cool water is good for a delicate molecule such as RNA, but hydrothermal events in the lake emit streams of hot water, which could provide transient exposure to the high temperatures needed for strand separation following template copying.

life began was probably cool most of the time, because RNA is a delicate molecule that will degrade if it is heated for too long. On the other hand, short periods of high temperature might have been essential so that once an RNA template had been copied, making a double-helical product, the complementary strands could be separated and copied again. Geothermally active areas can provide such a thermally fluctuating environment, making volcanically active areas or impact crater lakes seem ideal. For example, in Yellowstone Lake there are hydrothermal vents that emit plumes of hot water (Figure 13), and one can imagine primitive cells being caught up in these plumes, heated quickly and then quickly cooled back down as they mix with the surrounding cold lake water. Alternatively, the early earth was bombarded by many asteroids and comets that would create large crater lakes. Such lakes are also hydrothermally active for thousands of years after the impact, and therefore provide another environment that could have played a role in nurturing the beginnings of life.

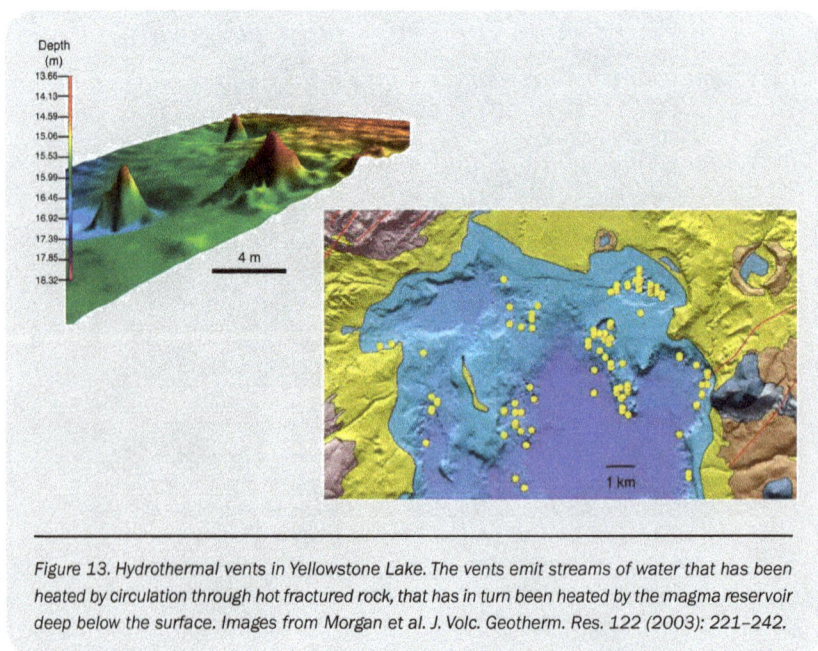

Figure 13. Hydrothermal vents in Yellowstone Lake. The vents emit streams of water that has been heated by circulation through hot fractured rock, that has in turn been heated by the magma reservoir deep below the surface. Images from Morgan et al. J. Volc. Geotherm. Res. 122 (2003): 221–242.

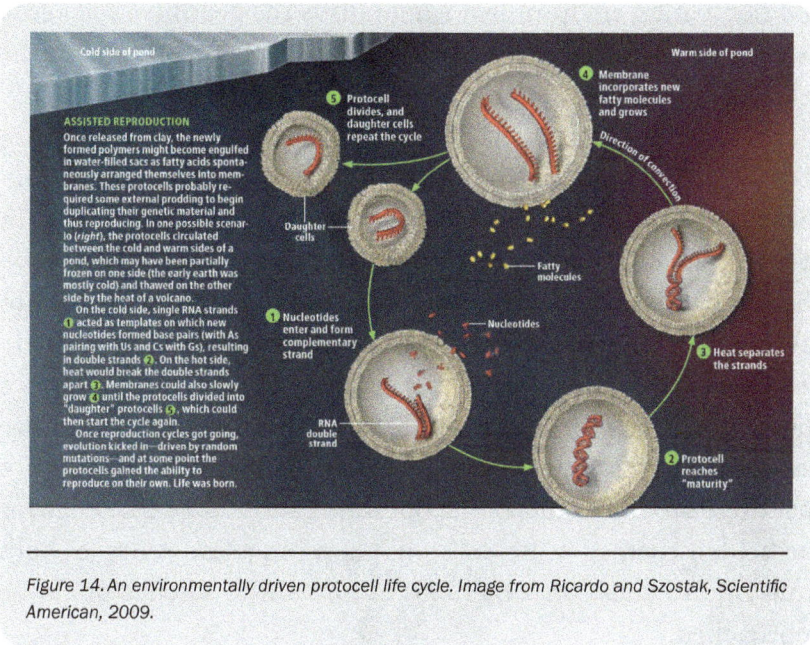

Figure 14. An environmentally driven protocell life cycle. Image from Ricardo and Szostak, Scientific American, 2009.

Because the first cells lacked internal biochemical machinery, it is likely that fluctuations in the environment controlled the processes of protocell growth, division, and replication (Figure 14)[29]. Imagine protocells living in a cold water environment that is chemically rich and contains a concentrated mix of all the necessary building blocks of biology (nucleotides, peptides, lipids and chemical sources of energy). Under such conditions, RNA copying could proceed along with growth of the protocell membrane. Every now and then these primordial cells would be pulled into a plume of hot water emanating from a hydrothermal vent, and the brief pulse of high temperature would serve to separate the strands of the RNA duplexes, while at the same time allowing for a sudden influx of nutrients to the cell interior. Before the high temperature could do any damage, the protocells would be swept back out into the cold lake waters, ready to go through another round of genome replication and membrane growth. In this scenario, a complex and fluctuating environment would drive the primitive cell cycle, before the evolution of internal machinery that could direct the cell cycle from inside.

I have presented an overview of where we are in our thinking about how life got started on the early earth. But, thinking about that kind of problem also raises other interesting questions. In everything I've discussed so far we have been trying to address the chemical and physical processes that would eventually give rise to life as we know it, in the form of modern biology. But could there be other possibilities? Would life emerge in a different form in a slightly, or in a very, different chemical environment? Could we design new kinds of life where the biochemistry is different? I think that these are really interesting questions and a huge challenge for the field of chemistry. Below I will point out a few examples both from my lab and from other labs of how this challenge might be approached.

As long as we are considering cellular life, the questions around how different such life could be reduce to the questions of whether it is possible to make membranes and genetic materials from non-biological building blocks, through non-biological chemistry? A good example of artificial membranes that are chemically distinct from biological membranes but still biophysically similar comes from the Devaraj lab at UCSD[30]. They have taken advantage of the very widely used copper click chemistry in which an azide and an alkyne can be joined together with a copper catalyst. This can be used to make lipids that look similar to biological phospholipids, but are chemically a little bit different and are made in a very different way. Remarkably, they have developed a catalyst for this lipid coupling reaction that can also catalyze its own synthesis. As a result, these membrane systems can grow indefinitely as long as they are fed with the correct synthetic building blocks. This is a potential route to making a non-biological protocell membrane system that could grow and divide indefinitely given the necessary resources.

What about a non-biological genetic material? One of the molecules related to RNA and DNA that we've been working on in my lab for quite a while is called 3'-NP-DNA or phosphoramidate DNA. The chemical structure of this polymer is the same as that of DNA except that the oxygen atom at the 3'-position of every sugar has been replaced by a nitrogen atom; in the corresponding nucleotide building blocks the 3'-hydroxyl group has been changed to a 3'-amine. The advantage of this change is that in the nucleotides are much more reactive, because an amine is a much stronger nucleophile than a hydroxyl. As a result these molecules are actually easier to copy in a purely chemical system without enzymes[31]. We are continuing to work on NP-DNA because, even though we are not yet able to fully replicate NP-DNA oligonucleotides, we are close enough that we can start to imagine the possibility of building living cells that use a genetic material that is different from RNA and DNA.

The examples of non-biological membranes and genetic polymers discussed so far remain quite close to biology as we know it, which raises the question of whether we can go further afield? Can we imagine making new kinds of living systems where the chemistry is really completely different? Speculation about such highly divergent forms of life was greatly stimulated when the Cassini-Huygens mission discovered the lakes and seas on Saturn's moon Titan (Figure 15). These are not lakes of water, but of liquid methane and ethane. Can we imagine anything living in an environment like that? Liquid methane is hard to work with in the lab, but we can easily work with nonpolar organic solvents such as decane. The challenge of making membrane vesicles in decane was actually addressed by the Kunieda lab in Japan, more than 20 years ago[32]. Figure 16 shows vesicles composed of inside-out membranes with the polar parts of the lipids in the membrane interior, and the hydrophobic parts sticking into the hydrophobic solvent. They look like completely normal vesicles but they're inside out and the molecular components are totally different. Given that

Figure 15. Lakes of methane and ethane on the surface of Saturn's moon, Titan. Could there be life in such environments?

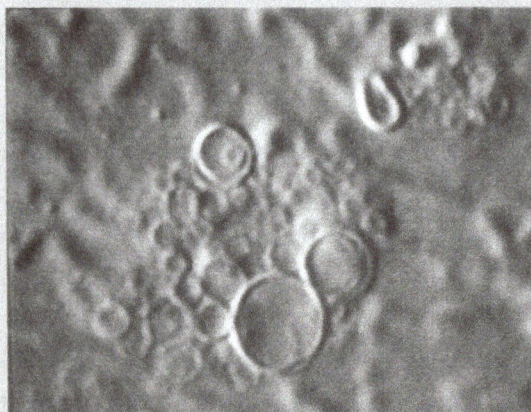

Figure 16. Reverse vesicles in decane. The vesicle membranes are assembled from molecules that contain hydrophobic regions that point out to the decane phase, and hydrophilic regions that cluster together in the middle of the membrane. Image from H Kunieda, K Nakamura, and DF Evans. Formation of Reversed Vesicles. JACS 1991 113: 1051–1052.

it is possible to make membranes in such a solvent, what kind of genetic material can we imagine in such a solvent? This is probably a more difficult challenge, but one that is almost irresistible. Our work on this is at a very early stage, but it is a fascinating project. One of the wonderful things about working on such different genetic polymers is that it makes us appreciate RNA and DNA even more, since we can see how hard it is to design something new that has the same properties.

To summarize, I have tried to present an overview of the new questions and new ideas that are changing how we think about the chemistry that led to life on the early Earth. I have also tried to point out some of the tantalizing beginnings of work that may lead to the ability to construct living systems that are chemically completely different from life on Earth. These are both fantastically interesting challenges, so stay tuned for future progress!

Acknowledgements

I would like to thank the many fantastic students, post-docs and collaborators that I have had the pleasure of working with over the years. Work in my lab on the origin of life has been supported by grants from the Simons Foundation (290363), the NSF (CHE-1607034) and NASA (NNX15AL18G). I am an Investigator of the Howard Hughes Medical Institute.

References

1. Lorsch, J.R. & Szostak, J.W. Chance and necessity in the selection of nucleic acid catalysts. *Accounts Chem. Res.* **29**, 103–110 (1996).
2. Szostak, J.W., Bartel, D.P. & Luisi, P.L. Synthesizing life. *Nature* **409**, 387–390 (2001).
3. Szostak, J.W. The eightfold path to non-enzymatic RNA replication. *J. System Chem.* **3**, 2 (2012).
4. Gillon, M., Triaud, A.H., Demory, B.O., Jehin, E., Agol, E., Deck, K.M., Lederer, S.M., de Wit, J., Burdanov, A., Ingalls, J.G., Bolmont, E., Leconte, J., Raymond, S.N., Selsis, F., Turbet, M., Barkaoui, K., Burgasser, A., Burleigh, M.R.,

Carey, S.J., Chaushev, A., Copperwheat, C.M., Delrez, L., Fernandes, C.S., Holdsworth, D.L., Kotze, E.J., Van Grootel, V., Almleaky, Y., Benkhaldoun, Z., Magain, P. & Queloz, D. Seven temperate terrestrial planets around the nearby ultracool dwarf star TRAPPIST-1. *Nature* **542**(7642), 456–460 (2017).

5. Sutherland, J.D. The origin of life — Out of the blue. *Angew. Chem. Int. Ed. Engl.* **55**(1), 104–121 (2016).

6. Miller, S.L. A production of amino acids under possible primitive earth conditions. *Science* **117**(3046), 528–529 (1953).

7. Orgel, L.E. Prebiotic chemistry and the origin of the RNA world. *Crit. Rev. Biochem. Mol. Biol.* **39**(2), 99–123 (2004).

8. Islam, S. & Powner, M.W. Prebiotic systems chemistry: Complexity overcoming clutter. *Chem.* **2**(4), 470–501 (2017).

9. Patel, B.H., Percivalle, C., Ritson, D.J., Duffy, C.D. & Sutherland, J.D. Common origins of RNA, protein and lipid precursors in a cyanosulfidic protometabolism. *Nat. Chem.* **7**(4), 301–307 (2015).

10. Ritson, D.J. & Sutherland, J.D. Synthesis of aldehydic ribonucleotide and amino acid precursors by photoredox chemistry. *Angew. Chem. Int. Ed. Engl.* **52**(22), 5845–5847 (2013).

11. Powner, M.W., Gerland, B. & Sutherland, J.D. Synthesis of activated pyrimidine ribonucleotides in prebiotically plausible conditions. *Nature* **459**(7244), 239–242 (2009).

12. Springsteen, G. & Joyce, G.F. Selective derivatization and sequestration of ribose from a prebiotic mix. *J. Am. Chem. Soc.* **126**(31), 9578–9583 (2004).

13. Hein, J.E., Tse, E. & Blackmond, D.G. A route to enantiopure RNA precursors from nearly racemic starting materials. *Nat. Chem.* **3**(9), 704–706 (2011).

14. Mansy, S.S., Schrum, J.P., Krishnamurthy, M., Tobé, S., Treco, D.A. & Szostak, J.W. Template-directed synthesis of a genetic polymer in a model protocell. *Nature* **454**(7200), 122–125 (2008).

15. Budin, I. & Szostak, J.W. Expanding roles for diverse physical phenomena during the origin of life. *Annu. Rev. Biophys.* **39**, 245–263 (2010).

16. Zhu, T.F. & Szostak, J.W. Coupled growth and division of model protocell membranes. *J. Am. Chem. Soc.* **131**(15), 5705–5713 (2009).

17. Zhu, T.F., Adamala, K., Zhang, N. & Szostak, J.W. Photochemically driven redox chemistry induces protocell membrane pearling and division. *Proc. Natl. Acad. Sci. U.S.A.* **109**(25), 9828–9832 (2012).

18. Nissen, P., Hansen, J., Ban, N., Moore, P.B. & Steitz, T.A. The structural basis of ribosome activity in peptide bond synthesis. *Science* **289**(5481), 920–930 (2000).

19. Inoue, T. & Orgel, L.E. Oligomerization of (guanosine 5'-phosphor)-2-methylimidazolide on poly(C). An RNA polymerase model. *J. Mol. Biol.* **162**(1), 201–217 (1982).

20. Szostak, J.W. The narrow road to the deep past: In search of the chemistry of the origin of life. *Angew. Chem. Int. Ed. Engl.* **56**(37), 11037–11043 (2017).

21. Li, L., Prywes, N., Tam, C.P., O'Flaherty, D.K., Lelyveld, V.S., Izgu, E.C., Pal, A. & Szostak, J.W. Enhanced nonenzymatic RNA copying with 2-aminoimidazole activated nucleotides. *J. Am. Chem. Soc.* **139**(5), 1810–1813 (2017).

22. Fahrenbach, A.C., Giurgiu, C., Tam, C.P., Li, L., Hongo, Y., Aono, M. & Szostak, J.W. Common and potentially prebiotic origin for precursors of nucleotide synthesis and activation. *J. Am. Chem. Soc.* **139**(26), 8780–8783 (2017).

23. Islam, S., Bučar, D.-K. & Powner, M.W. Prebiotic selection and assembly of proteinogenic amino acids and natural nucleotides from complex mixtures. *Nat. Chem.* **9**, 584–589 (2017).

24. Engelhart, A.E., Powner, M.W. & Szostak, J.W. Functional RNAs exhibit tolerance for non-heritable 2'-5' versus 3'-5' backbone heterogeneity. *Nat. Chem.* **5**(5), 390–394 (2013).

25. Heuberger, B.D., Pal, A., Del Frate, F., Topkar, V.V. & Szostak, J.W. Replacing uridine with 2-thiouridine enhances the rate and fidelity of nonenzymatic RNA primer extension. *J. Am. Chem. Soc.* **137**(7), 2769–2775 (2015).

26. Xu, J., Tsanakopoulou, M., Magnani, C.J., Szabla, R., Šponer, J.E., Šponer, J., Góra, R.W. & Sutherland, J.D. A prebiotically plausible synthesis of pyrimidine β-ribonucleosides and their phosphate derivatives involving photoanomerization. *Nat. Chem.* **9**(4), 303–309 (2017).

27. Leman, L., Orgel, L. & Ghadiri, M.R. Carbonyl sulfide-mediated prebiotic formation of peptides. *Science* **306**(5694), 283–286 (2004).

28. Izgu, E.C., Björkbom, A., Kamat, N.P., Lelyveld, V.S., Zhang, W., Jia, T.Z. & Szostak, J.W. N-carboxyanhydride-mediated fatty acylation of amino acids and peptides for functionalization of protocell membranes. *J. Am. Chem. Soc.* **138**(51), 16669–16676 (2016).

29. Ricardo, A. & Szostak, J.W. Origin of life on earth. *Sci. Am.* **301**(3), 54–61 (2009).

30. Hardy, M.D., Yang, J., Selimkhanov, J., Cole, C.M., Tsimring, L.S. & Devaraj, N.K. Self-reproducing catalyst drives repeated phospholipid synthesis and membrane growth. *Proc. Natl. Acad. Sci. U.S.A.* **112**(27), 8187–8192 (2015).

31. Zhang, S., Blain, J.C., Zielinska, D., Gryaznov, S.M. & Szostak, J.W. Fast and accurate nonenzymatic copying of an RNA-like synthetic genetic polymer. *Proc. Natl. Acad. Sci. U.S.A.* **110**(44), 17732–17737 (2013).

32. Kunieda, H., Nakamura, K. & Evans, D.F. Formation of reversed vesicles. *J. Am. Chem. Soc.* **113**, 1051–1052 (1991).

Making Materials That Hate Water to Love Water: The Transformative Power of Chemistry*

Richard N. Zare[1]

A simple nonaqueous reaction scheme for transforming the surface of plastics from hydrophobic to hydrophilic is presented. The chemical modification is achieved by the base-catalyzed trans-esterification of polyethylene terephthalate (PET), which is a commonly used plastic. Its use in blood collection tubes is discussed. The surface modification is permanent, inexpensive, rapid, and does not release contaminants. It also causes no optical or mechanical distortion of the plastic. This work demonstrates the power of chemistry to transform common materials.

Keywords: Hydrophilic; Hydrophobic; Polyethylene Terephthalate.

---

* This chapter was based on a lecture delivered by Professor Zare at the Molecular Frontiers Symposium in Chalmers University of Technology, Gothenburg in November 2014.
[1] Department of Chemistry, Stanford University, Stanford, California, USA.

About Richard Zare

Richard N. Zare is the Marguerite Blake Wilbur Professor in Natural Science at Stanford University. He was born on November 19, 1939 in Cleveland, Ohio, and is a graduate of Harvard University, where he received his B.A. degree in chemistry and physics in 1961 and his Ph.D. in chemical physics in 1964. In 1965 he became an assistant professor at the Massachusetts Institute of Technology, but moved to the University of Colorado in 1966, remaining there until 1969 while holding joint appointments in the departments of chemistry, and physics and astrophysics. In 1969 he was appointed to a full professorship in the chemistry department at Columbia University, becoming the Higgins Professor of Natural Science in 1975. In 1977 he moved to Stanford University. He was named Chair of the Department of Chemistry at Stanford University in 2005. In 2006 he was named a Howard Hughes Medical Institute (HHMI) Professor.

Zare is renowned for his research in the area of laser chemistry, resulting in a greater understanding of chemical reactions at the molecular level. By experimental and theoretical studies he has made seminal contributions to our knowledge of molecular collision processes and contributed very significantly to solving a variety of problems in chemical analysis. His development of laser induced fluorescence as a method for studying reaction dynamics has been widely adopted in other laboratories.

Zare has received numerous honors and awards. They include: Phi Lambda Upsilon's Fresenius Award (1974), Michael Polanyi Medal, the Gas Kinetics Group of the Royal Society of Chemistry (1979), the APS Earle K. Plyler Prize (1981), Spectroscopy Society of Pittsburgh Award (1983), the National Medal of Science (1983), the Evans Award and Lectureship, Department of Chemistry, The Ohio State University (1984), the ACS (Maryland Section) Remsen Award (1985), the ACS (Rochester Section) Harrison Howe Award (1985), the APS Irving Langmuir Prize (1985), the ACS (New Haven Section) Kirkwood Medal (1986), Michelson-Morley Award, Case Western Reserve University (1986), the ACS (Chicago Section) Willard Gibbs Medal (1990), the ISCO Award for Significant Contributions to Instrumentation for Biochemical Separations (1990), The National Academy of Sciences Award in Chemical Sciences (1991), the ACS Peter Debye Award in Physical Chemistry (1991),

The Harvey Prize (1993), the Dannie-Heineman Preis (1993), the ACS (Puget Sound, Oregon and Portland Sections) Pauling Award (1993), the ACS (Division of Analytical Chemistry) Award in Chemical Instrumentation (1995), NASA Exceptional Scientific Achievement Award (1997), the California Scientist of the Year Award (1997), the Eastern Analytical Symposium Award for Outstanding Achievements in the Field of Analytical Chemistry (1997), National Science Board's Distinguished Service Award (1998), the ACS (Auburn Section) G. M. Kosalapoff Award (1998), the ACS Award in Analytical Chemistry (1998), the Centennial Medal, Graduate School of Arts and Sciences, Harvard University (1998), the ACS E. Bright Wilson Award in Spectroscopy (1999), the Welch Award in Chemistry (1999), the APS Arthur L. Schawlow Prize in Laser Science (2000), the ACS (North Alabama Section) Madison Marshall Award (2000), the California Separation Science Society Scientific Achievement Award (2000), the ACS Nobel Laureate Signature Award for Graduate Education (2000), Royal Society of Chemistry Faraday Lecture (2001), the ACS Charles Lathrop Parsons Award (2001), the ACS (Sierra Nevada Section) Distinguished Chemist Award (2002), the ACS (New York Section) Nichols Medal (2004), the Chandler Medal, Department of Chemistry, Columbia University (2005), Pupin Medal "for service to the nation," Columbia University School of Engineering (2005), the Wolf Prize in Chemistry, Israel (2005), the ACS (University of Cincinnati and Cincinnati Section) Oesper Award (2006), the Dudley R. Herschbach Award for Excellence in Research in the field of Collision Dynamics, Dynamics of Molecular Collisions Meeting, Santa Fe (2007), the H. Julian Allen Award, NASA Ames Research Center (2007), the Texas A&M University, Department of Chemistry, and Texas A&M Section, ACS, F. A. Cotton Medal for Excellence in Chemical Research (2009), the ACS (Northeastern Section) Richards Medal (2010), the ACS Priestley Medal (2010), the BBVA Foundation Frontiers of Knowledge Award in the Basic Sciences category (shared with Michael E. Fisher) (2010), the Honorary Fellow of the Chinese Chemical Society (elected 2010), the R. B. Bernstein Award in Stereodynamics (shared with R.D. Levine) (2010), the King Faisal International Prize in Science (shared with George M. Whitesides) (2011), Honorary Membership into the Japan Society for Analytical Chemistry (2011), the Einstein Professorship of the Chinese Academy of Sciences (2011), the Torbern Bergman Medal (2012), the International Science and Technology Cooperation Award of the People's Republic of China (2012), and the Othmer Gold Medal from the Chemical Heritage Foundation (2017).

Teaching awards: Stanford University's Dean's Award for Excellence in Teaching (1987), the Bing Fellowship Award, "to recognize excellence in teaching and commitment to undergraduate education," Stanford University (1996), the Allan V. Cox Medal for Faculty Excellence Fostering Undergraduate Research, Stanford University (1997), the Laurance and Naomi Carpenter Hoagland Prize for excellence in undergraduate education, Stanford University (2003), the ACS (Northeastern

Section) James Flack Norris Award for Outstanding Achievement in the Teaching of Chemistry (2004), the ACS George C. Pimentel Award in Chemical Education (2008), and the Presidential Award for Excellence in Science, Mathematics, and Engineering Mentoring (2009).

Zare holds honorary degrees from: the University of Arizona (1990), Northwestern University (1993), Eidgenössische Technische Hochschule Zürich (1993), Uppsala University (2000), Columbia University (2000), The University of York (2001), The State University of West Georgia (2001), Hunan University (2002), Université Paul Sabatier (2003), Chalmers Institute of Technology (2007), the University of South Florida (2013), and the University of Edinburgh (2016).

Member of the the National Academy of Sciences (1976), the American Academy of Arts and Sciences (1976), American Philosophical Society (1991), and the World Jewish Academy of Sciences (2004). Fellow of the American Physical Society (1969), American Association for the Advancement of Science (1985), the Association for Women in Science (2008), and the American Chemical Society (2009). Honorary Fellow of California Academy of Sciences (1991), the Royal Society of Chemistry (2001), the Chemical Research Society of India (CRSI) (2007), and the Indian Academy of Sciences (2008). Fellow of TWAS, Academy of Sciences for the Developing World (2009). Non-Resident Fellow of the Joint Institute of Laboratory Astrophysics (1969). Foreign Member of the Royal Society (London) (1999), the Chinese Academy of Sciences (2004), and the Swedish Royal Academy of Engineering Sciences (2004).

During 1992–1995 Zare chaired the National Research Council's Commission on Physical Sciences, Mathematics, and Applications, and chaired the National Science Board the last two years of his 1992–1998 service. In 1997–2000 Zare served as the Chair of the President's National Medal of Science Selection Committee. In 2007, he was appointed to be an advisor to the Camille and Henry Dreyfus Foundation and became a member of the Board of Directors in 2010. And most recently in 2012 Zare was appointed chair of the Committee on Science, Engineering and Public Policy (COSEPUP) of the three academies, the National Academy of Sciences, the National Academy of Engineering, and the Institute of Medicine. He also currently acts as Chairman of the Board of Directors at Annual Reviews, Inc.

Zare has given named lectures at numerous universities, has authored and co-authored over 900 publications and more than 50 patents, and he has published four books.

What a pleasure to be with you this morning and to have this opportunity to talk to you about materials for tomorrow, really something about the transformative power of chemistry. And I'm going to pick one particular material, one that I hope many of you are familiar with. I'm going to talk about blood collection tubes. At first, this sounds not very exciting. But consider how many people have ever had their blood taken. I can tell you in the United States — I only have the figures for the United States — 3 billion tubes are made each year in which we collect blood and analyze it. So it is one of the major ways of being medically examined. Yet, the story behind blood collection tubes has many twists and turns, as I will explain. Here is the story of how I became involved.

Dr. Raffick Bowen (Figure 1), who is head of the Stanford Hospital in-charge of drawing and analyzing blood, came to me and he told me "I got a problem. Help, can you do something? I used to be able to use glass for my blood collection tubes but I am no longer allowed to do that. Doctors and nurses keep dropping the glass blood collection tubes on the floor and infecting themselves. So now I am only permitted to use plastic, and the plastic I use is

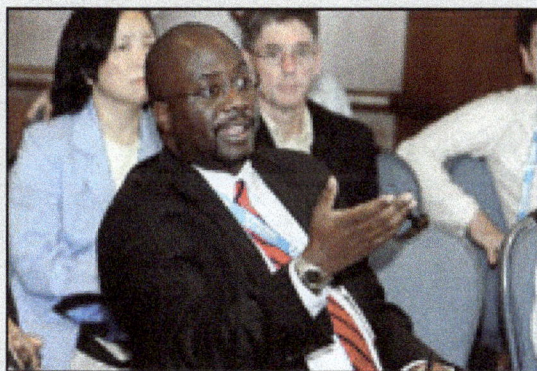

Figure 1. Dr. Raffick A. R. Bowen, Department of Pathology, Stanford University School of Medicine.

made of something called polyethylene terephthalate also simply called PET. But the plastic pulls molecules out of our blood samples and interferes with our blood analysis. Can't you make the plastic act like glass?"

I was surprised I was asked as I am not known for my prowess in chemical synthesis. I did have a reputation of being willing to pursue problems outside my field of expertise. I did remember some organic chemistry from long ago, and I was able to convince a graduate student of mine, Samuel L. Kim (Figure 2), to give this problem a try, and very quickly something worked.

I'm going to tell you about how to change the surface of blood collection tubes that are plastic, from hating water to loving water, that is, to transform the surface from hydrophobic to hydrophilic. I'll go through how this is done.

Let's begin with the importance of this topic. It turns out blood collection and processing are the two major pre-analytical steps in clinical laboratory testing. If you look into where the errors in blood analysis occur, often it is caused by the blood collection tube, as revealed by a brief history of this topic. At first, people started using reusable glass syringes with steel hypodermic needles. Those were the

Figure 2. Dr. Samuel L. Kim, my former graduate student, who is presently at UCSF Medical School.

first devices. And the glass syringes were expensive to manufacture and they could easily break. What ultimately prompted the replacement, however, was not that fact, but rather something which I think will be obvious to everybody. You don't want to keep using the same needle in people. And this led very much to outbreaks of hepatitis. Since the 1940s, people have used evacuated blood collection tubes because they could automatically draw a predetermined blood volume and switching between tubes for additional samples pose a lower risk of spillage and needle stick injury and you use it all at once and throw it away. That's been the hope of this. So far good.

Glass evacuated tubes containing anticoagulants were commonly used from the 1950s to the 1990s. Anticoagulants stopped the blood from coagulating, from thickening. You think this is easy. It's not easy. They started by using the wrong form of glass. They used something called soda lime glass and it would actually leach out things like calcium ions that would interfere with the tests they were doing. So they had to stop that. And then another thing happened. And this was more recent, at least in the United States. An organization called OSHA decided that it was unsafe to use glass in hospitals and clinics. Why? Because doctors and nurses would drop the blood-filled glass tubes; they would shatter, and then they would cut themselves on the glass and could become infected. So now in the USA OSHA has demanded that blood collection tubes be made from plastic. Plastic doesn't break when you drop it. That's of course good — like plastic milk bottles. The same switch has been taking place in many places. So that's what's happening.

They're manufactured through an injection molding using what are called polyesters. I'll show you what those are. These are polymers, most commonly being something called polyethylene terephthalate or PET. You've seen lots of bottles with PET as its name or combinations like that. And now what is it? Figure 3 shows the chemical structure of PET.

There is a linkage called a polyester linkage that connects the rigid terephthalate group with the flexible ethylene group. This is

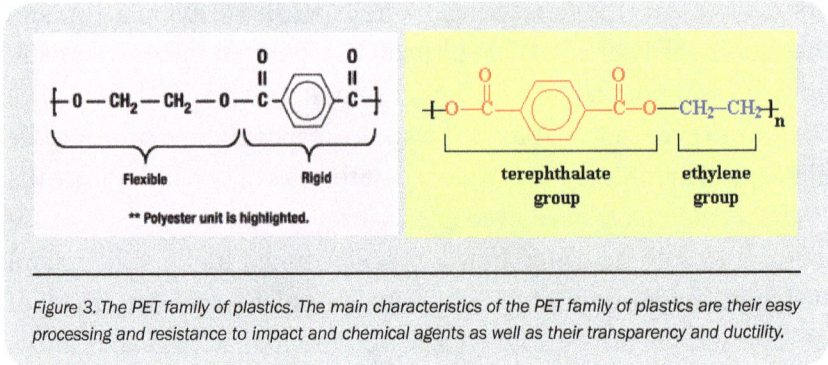

Figure 3. The PET family of plastics. The main characteristics of the PET family of plastics are their easy processing and resistance to impact and chemical agents as well as their transparency and ductility.

where an acid and the base would react to form an ester. Notice that this motif is repeated again and again; that's how you make polymers.

Polymers are an amazing thing. If it weren't for polymers, most of us would lose our clothes and be a bit embarrassed. So much is done with polymers today, including these tubes. So here's what's good about them. Compared to glass, plastic minimizes exposure to biohazardous materials following breakage. It has a greater shock resistance, it tolerates higher centrifugation speeds, people spin it around, centrifuge it. The plastic tube weighs less and has excellent dimensional precision, and is more easily disposed of by incineration at a lower cost. It is very hard to incinerate glass. This part works well.

Let me give you next a sense of what's bad, as pictured in Figure 4. Here you see water balling up on this leaf. This is because the leaf is waxy and is what we call hydrophobic, that is, hates water. And plastic tubes generally have a hydrophobic surface. They hate water and they do not efficiently activate the coagulation process. Clots formed on the plastic surface of tubes are gelatinous compared to clots formed in glass tubes. Blood does not flow smoothly. No one today uses bare plastic tubes. They are that bad: you have to fix this problem. They take out things from the blood and interfere with the tests.

So what does the manufacturer of blood collection tubes do? The fix to this problem is you spritz in, you squirt in, detergent; you add

Figure 4. The surface of a leaf with morning dew condensed on it and balling up into droplets.

a surfactant coating to the inside of the blood collection tube. But these surfactants can leach out and interfere with the blood analysis. And this is why Dr. Bowen came to see me because they're causing him a lot of trouble. He says do something about this.

I'm going to show you what I think is relatively easy chemistry; in some sense, it looks like magic, but it's something that all of you could do if you want to and it shows you the power of chemistry. Here we have a surface that hates water. How are we going to make it so it likes water? It's a common problem. It's a problem that we all face about what causes things to dissolve other things. And generally the expression is that like likes like. You want to make something similar. So that to make it similar what we really want to do is add –OH groups. You all know that water is H_2O, that is, H–OH. Well, you want to put on the outside of the surface –OH groups called hydroxyl groups. That's what we want to do. And I'm going to show you how we do that.

Here is then a little bit of some chemistry, as shown in Figure 5. In organic chemistry, a nucleophilic addition reaction is an addition reaction where a chemical compound with an electron deficient or electrophilic double or triple bond, a π bond, reacts with an electron rich reactant, termed a nucleophile. The magic word in chemistry is

Figure 5. An alcohol R'OH is added to an ester R"O(C=O)R to cause R' and R" to be interchanged. Strong acids catalyze the reaction by donating a proton to the C=O carbonyl group, thus making it a more potent electrophile. Strong bases catalyze the reaction by removing a proton from the alcohol, thus making it more nucleophilic. In what follows we use the latter approach.

catalysis — to find something to speed things up which itself doesn't really play a role in what's happening. It intimately gets involved but it's recovered, it's recycled in the process. Look what happens. If you add, however, a strong acid to an alcohol, you're going to form an ester. This is not a good idea, this is the wrong approach. Instead a strong base is what you want and you want a strong base to catalyze the reaction by removing a proton from the alcohol and thus making it what we would call nucleophilic which makes if a better attacking group.

Now comes the next idea and it's an idea that's taken me a while to first realize. I didn't get this at all in high school. It just came up later. The idea is that you can actually find a base stronger than OH⁻, but the only way you can do that is not to work in water. Water, which is 55 molar, stops you. You must go to what's called a nonaqueous medium; then you can work with things called superbases. And superbases are extremely strong bases that have a higher affinity for protons than the OH⁻ group. The hydroxide ion is the strongest base possible when you have water but bases exist with even much greater strength for protons and we're going to use these.

So here's our proposed fix. We plan to chemically convert the inner surface of the PET blood collection tube from hydrophobic to hydrophilic by adding an alcohol and a superbase in the absence of water. That's the goal I had in mind. Figure 6 shows you what we do.

The particular alcohol we added is called ethylene glycol, which is the common constituent of anti-freeze. At first, Sam Kim tried

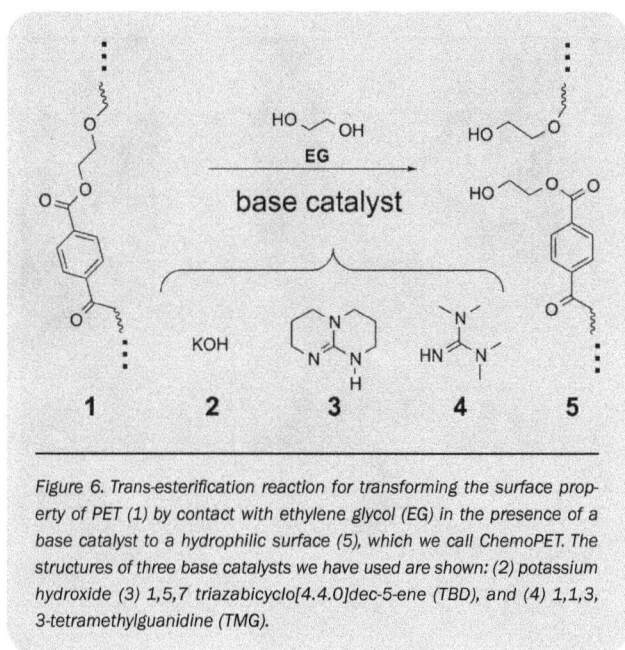

Figure 6. Trans-esterification reaction for transforming the surface property of PET (1) by contact with ethylene glycol (EG) in the presence of a base catalyst to a hydrophilic surface (5), which we call ChemoPET. The structures of three base catalysts we have used are shown: (2) potassium hydroxide (3) 1,5,7 triazabicyclo[4.4.0]dec-5-ene (TBD), and (4) 1,1,3, 3-tetramethylguanidine (TMG).

potassium hydroxide but later on we replaced it with superbases. They have names you can read in the caption to Figure 6 but I'm not going to bother to say them. In this trans-esterification process of PET we put –OH groups on the outside of the inner surface of the plastic blood collection tube. That's what we're doing. And that could be really neat. It turns out the chemistry is not original. It's been done before. Believe me it is not original at all. But the idea of making a new material, particularly a surface, appears to be original, and we have actually filed for a patent.

So the question is of after this big preamble, does it work? Well, would I stand here before you if I did have some idea that it worked, but let me try to convince you that it does work. Perhaps, the best way to do that is to show you two test tubes made out of plastic, made out of PET (Figure 7).

Look at the PET blood collection tube that's unmodified. The water just doesn't like its hydrophobic surface and the meniscus is

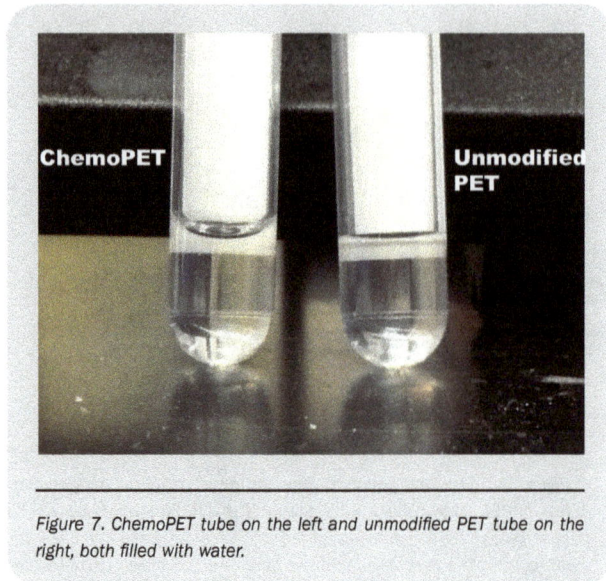

Figure 7. ChemoPET tube on the left and unmodified PET tube on the right, both filled with water.

flat. Contrast this behavior with the ChemoPET blood collection tube whose surface has been made hydrophilic. The water meniscus bends upward because the water loves the hydrophilic surface. The surface of the ChemoPET blood collection tube is acting like the surface of a glass blood collection tube!

And I'm showing you before treatment and after treatment and I'm doing it on plaques of PET where you can actually again see how this water droplets spreads out. And it looks good, as shown in Figure 8.

What further excites me about this treatment process is it can be done in about ten minutes at room temperature, with essentially no loss of catalyst and hardly any consumption of the ethylene glycol because this is a surface modification. It also has an outstanding shelf life. Consequently, the making of ChemoPET tubes from PET tubes looks to be inexpensive, quick, and not very laborious. Only time will tell whether this can supplant existing blood collection tubes made by an

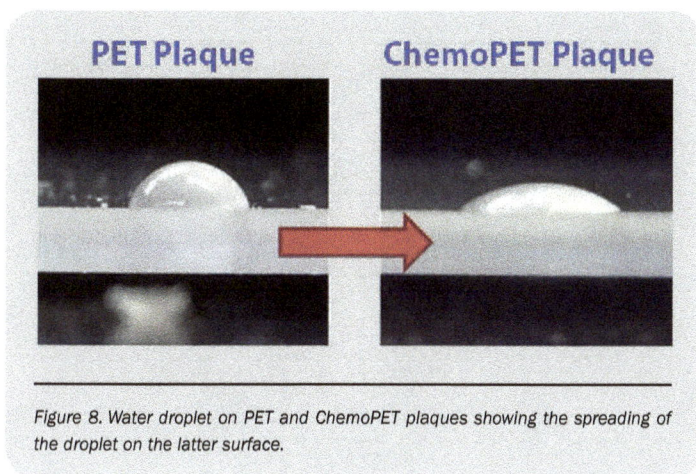

Figure 8. Water droplet on PET and ChemoPET plaques showing the spreading of the droplet on the latter surface.

established and entrenched technology. We have been assessing its performance in blood collection analysis and it seems to be comparable to the use of glass blood collection tubes. More detailed comparisons have been published after this presentation was made: (1) Kim, S., Bowen, R.A.R. & Zare, R.N. Transforming plastic surfaces with electrophilic backbones from hydrophobic to hydrophilic. *ACS Appl. Mater. Interfaces* 7, 1925–1931 (2015). (2) Bowen, R.A.R., Kim, S.C., Sattayapiwat, A., Austria-Esguerra, V. & Zare, R.N. Performance of chemically modified plastic blood collection tubes. *Clin. Biochem.* **49**, 90–99 (2016).

I hope this short presentation has given you an idea of this research field. It's one example of the power of chemical transformations and it shows what you're able to do with a little bit of chemistry understanding to really change the nature of materials. Yes, it's magic. But it's magic that we understand and can control and do much more with. There's going to be yet more exciting developments, I believe, presented to you today that will impact our lives. Here's a very simple one. And I thank you to have this opportunity to present this to you.

www.ingramcontent.com/pod-product-compliance
Lightning Source LLC
Chambersburg PA
CBHW050545190326
41458CB00007B/1931